THE CASE AGAINST FRAGRANCE

Kate Grenville is one of Australia's most celebrated writers. Her international bestseller *The Secret River* was awarded local and overseas prizes, has been adapted for the stage and as a TV miniseries, and has become a much-loved classic. Grenville's other novels include *Sarah Thornhill*, *The Lieutenant*, *Dark Places* and the Orange Prize winner *The Idea of Perfection*. Her most recent book is the acclaimed *One Life: My Mother's Story*.

kategrenville.com

THE CASE
AGAINST
FRAGRANCE

KATE
GRENVILLE

TEXT PUBLISHING MELBOURNE AUSTRALIA

textpublishing.com.au

The Text Publishing Company
Swann House
22 William Street
Melbourne Victoria 3000
Australia

First published by The Text Publishing Company 2017

Book design by Imogen Stubbs
Typeset by J&M Typesetting

Printed and bound in Australia by Griffin Press, an accredited ISO/NZS 1401:2004 Environmental Management System printer

National Library of Australia Cataloguing-in-Publication entry
ISBN: 9781925355956 (paperback)
ISBN: 9781925410310 (ebook)
Creator: Grenville, Kate, 1950– author.
Title: The case against fragrance / by Kate Grenville.
Subjects: Perfumes—Analysis. Perfumes industry. Odors—Analysis.
Dewey Number: 668.54

CONTENTS

'This is an era of specialists, each of whom sees his own problem and is unaware of or intolerant of the larger frame into which it fits. It is also an era dominated by industry, in which the right to make a dollar at whatever cost is seldom challenged...The public must decide whether it wishes to continue on the present road, and it can do so only when in full possession of the facts. In the words of Jean Rostand, "The obligation to endure gives us the right to know."'

RACHEL CARSON, *Silent Spring*

CHAPTER 1

PLANET FRAGRANCE

When I was little, my mother had a tiny, precious bottle of perfume on her dressing table and on special occasions she'd put a dab behind her ears. The smell of Arpège was always linked in my mind with excitement and pleasure—Mum with her hair done, wearing her best dress and her pearls, off for a night out with Dad.

Later, old enough to have my own special occasions, I also had my favourite perfume. I loved the bottles: those sensuous shapes. I loved the names and the labels, so evocative of all things glamorous. Above all, I loved what they promised: elegance, beauty, poise and—of course—desire. I'd set off for a date in a cloud of scent, feeling sumptuous and sexy. (The boys I was dating were enveloped in their own clouds of Old Spice, a smell forever associated for me with the slippery back seats of cars.) The boys never commented on my perfume, but one day a woman I envied for her smartness and stylishness exclaimed to me, 'Oh, you're wearing White Linen!' and I basked in her tone of respect and admiration.

I couldn't understand what happened a short time after I'd sailed out of the house in a cloud of scent. For half an hour I'd feel sumptuous and sexy, and then I'd get a headache. My eyes would get dry and sore. I'd get congested in the nose. I'd just want to go home. Was it the strain of being sumptuous and sexy? Did I have some sort of phobia about going on dates? If I thought about it at all, it was to blame myself.

Thinking back, it might even have been one of the things that made me a writer. You don't need to sit in a cloud of elegance when you're alone at the desk trying to think of another way of saying 'he said'.

Over time, the perfume bottle gravitated from the top of the chest of drawers into the undies-and-bras drawer. I never made a conscious decision, never really thought about it: but somehow, by the time my children came along in my thirties, I'd stopped using scent.

It was around then that a friend came back from Europe with a pretty little bottle of something. 'I thought this would suit you,' she said. 'I thought you'd like it.' I didn't know whether it suited me or not, but I did know that, as soon as I dabbed it on, a headache started up.

It was so immediate and so unexpected that it made me start to experiment. Finally, for the first time in all those years, I could see a pattern. When I used perfume, I got a headache. When I didn't, I didn't.

It was good to have that mystery solved: it wasn't dates I had a problem with, just perfume. That was all right. Some people couldn't bear the feel of wool against their skin, some people couldn't stand oysters: I was someone who didn't like perfume.

It seemed wasteful to throw away those pretty bottles with their promise of glamour, but once they were gone I never thought about them again.

As the years went by I came to dislike other kinds of perfume, too: the sickly fragrance in cosmetics, shampoo and cleaning products. By the time I was middle-aged I was looking for things that advertised themselves as 'fragrance-free' or 'no synthetic fragrance'. I had no problems with the scents of flowers, and small amounts of essential oils were okay, too. The problems seemed to be with the other sorts of smells. Whether you called the stuff fragrance, or scent, or perfume, and whether it came in an expensive little bottle with a French name, or as part of shampoo or soap, I avoided it when I could.

Then, in my fifties, I got a virus and took a long time to get better. Afterwards I realised that I'd become a lot more sensitive to scents. Now it wasn't just what I used myself—it was the stuff other people were using, too. For quite a time I tried to ignore an unwelcome fact: pretty much anything that involved being with other people in a confined space would mean a headache.

Then there was a night at the opera. I don't often go to the opera, but this was a special occasion, a night out with friends. It was going to be great: Teddy Tahu Rhodes being Don Giovanni. Good company, wonderful music, a great story. And at some point in the performance Teddy was sure to find an excuse to take off his shirt.

The first half was fabulous. At interval everyone filed out for their champagne but my friends and I stayed in our seats. Then everyone came back in and the lights went down for the second half. Within a few minutes I had a raging headache, my eyeballs felt too big for their sockets, and something had happened to my brain so I couldn't think, could barely hear the music, didn't care if Teddy took off the whole damn lot.

What had happened? Someone just behind me had taken advantage of the interval to refresh her perfume. Powerful blasts of it were rolling over me and everyone else nearby. The rest of the opera was a blur of headache and sore eyes, along with a strange fog in my brain that made everything feel far away and confused. All I wanted was to go home to bed.

That evening at the opera was the end of being an ostrich. I had to admit that something in my life had shifted.

My GP didn't think the headaches were anything serious, and the idea that perfumes and scented products might bring them on didn't seem to surprise her. A sensitivity, she called it. Not an allergy, but a sensitivity. Like being in a closed room with fresh paint.

Just to be sure, she sent me to a neurologist. He was a blunt man with nothing in the way of bedside manner and I expected him to scoff when I asked him whether the headaches could possibly have anything to do with fragrance or perfume. Oh yes, he said, that stuff was the very devil for giving people headaches. He reeled off the brand names of the half-dozen scents and colognes that in his experience were the worst offenders. But there was no specific test for fragrance sensitivity, he said, and no treatment other than avoidance.

—

Until it happened to me, I'd never heard of anyone who had a problem with fragrance. As far as I was concerned, the smells in perfume, cologne, and scented bathroom and cleaning products were some ingenious concoction of essence-of-flowers, put together by glamorous French people. My feeling was, if I kept cheerful and took a bit of evasive action, the fragrance sensitivity would probably go away, as the virus eventually had.

I wrote a light-hearted little piece for my website, 'Thanks for Not Wearing Fragrance', and developed a few tricks for avoiding the worst of other people's perfumes and the headaches that would follow. I chose outside tables in cafes. I fended off women who wanted to greet me with a kiss and when they said, 'Oh, got a cold?' I nodded. I asked my friends not to wear perfume when we saw each other, and being friends they agreed, and most of them remembered.

Then in early 2015 I published a book about my mother and went on tour to promote it. It was going to take me to every state over months of intermittent travelling, usually with Jane, the publicity manager at Text Publishing.

The moment I got into the cab to go to the airport I realised I was going to have a difficult time. A fragrance diffuser—a little glass thing full of liquid fragrance—was plugged into the air vent. Within half a minute the familiar headache was back. The airport was full of perfume shops. The woman next to me on the plane radiated something strong and musky. The cab at the other end had another fragrance diffuser.

The publishers had booked me into a lovely boutique hotel but, as soon as I pushed the heavy door, entered the tasteful foyer and approached the smiling desk clerks, my heart sank. Along with the tasteful furnishings, the subdued music and the gracious receptionists was a powerful flowery smell, wafting out from candles burning in various parts of the room. The lift was full of patchouli and all along the corridor to my room were those reeds in bottles, wafting out something musky. I got to the room at last, closed the door, took a deep breath of relief. But no, I could still smell fragrance. The hotel was air-conditioned and, because there were only a few rooms, the same perfumed air was circulating through the whole building. I went to the window, but it couldn't be opened.

It was a miserable night. In the morning, embarrassed but determined, I explained the problem to Jane. A good publicist takes the strangest requests in stride and, being the professional she is, she got straight on the phone and found a less grand but less odorous hotel.

The rest of the tour was a struggle against headache and brain fog. Until then, I hadn't realised just how hard it was to avoid fragrance in one form or another. But I learned cunning. I sprinted through the scented foyers of hotels with a scarf over my nose. In cabs I sat in the back, asked the driver to remove the fragrance diffuser, opened the window and stuck my head into the slipstream like a dog, arriving bedraggled and windblown. I carried unfragranced soap in my handbag for when I hadn't been quick enough to fend off a fragranced hug or a handshake, and after every event I retreated as quickly as I could to the hotel room to take off my fragrance-impregnated clothes and get under the shower.

Jane was always understanding, never made me feel neurotic or difficult. She was often the one who asked the cabbie to put the fragrance dispenser in the glove box. She went into the hotels first, leaving me outside while she did the paperwork, so I could make a quick dash through the fragranced foyer. She got me away from strongly perfumed women in the signing line and discouraged people from taking selfies too close to me.

The best part of book promotion is meeting readers. On this tour, waves of goodwill and warm feelings were pouring out from people—mostly women—who'd enjoyed the book I'd written about my mother. They had no idea in the world that a choice they'd made that morning was resulting in me having a headache. It seemed a situation with no solution. Fragrance in all its forms was an automatic part of their lives. And why shouldn't it be? It was I who was the problem. I felt embarrassed, somehow ashamed, and very alone.

The last hotel of the tour was in Launceston. Thank heavens, the room itself wasn't fragranced, and you could open the window. But the scent in the corridor was so strong that I could smell it pouring into the room through the crack between the door and the jamb.

I didn't unpack, just went straight back out into the street and walked until I found a newsagent where I could buy a roll of packaging tape. That night, when all the book business was done and I'd said goodnight to Jane, I taped up the crack around the door. I showered, washed the accumulated scents out of my hair, hung the fragranced clothes in the bathroom and shut the door on them. I lay on the bed, sealed in like a pharaoh, feeling the headache gradually fade.

As I lay admiring my handiwork, making a mental note always to travel with a roll of tape, I realised I wasn't going to tell Jane about what I'd done. I had a nasty feeling that I'd just crossed one of life's little boundaries. It was possible I'd joined the section of humanity that thinks the moon landings were faked by the CIA or the government puts Prozac in the water supply. After all, everyone else took fragrance for granted as a normal part of their lives. They weren't reduced to sealing themselves into hotel rooms with packaging tape.

Finally, I sat up on the gigantic bed and opened my laptop. When you google 'fragrance headache' you release a great chorus of voices. I discovered a thoroughly researched piece about fragrance by a science writer, Clare Pain, on the ABC website. The article, entitled 'When Others Abhor the Fragrance You Adore', was fascinating, but I was especially struck by the long string of posts in response, including these: 'If there is an aroma I just go home on sick leave.' 'Strong perfumes give me a severe sinus headache within minutes.' 'People don't seem to be able to understand how ill scents make me.' 'I get terrible migraines triggered by perfumes—deodorant sprays...are especially bad.' 'When I come in contact with perfumes I get a violent headache.'[1]

I stared at the car-sales yard out the window. This was

a surprise, and it was also a relief. I might be crazy, but I wasn't the only one.

That night I came across a reference to a study about fragrance, and I went looking for the original. It turned out to be a heavy-duty scientific paper, published in 2002 in the peer-reviewed, scholarly *Flavour and Fragrance Journal*. Its summary was so startling I had to read it a few times to make sure it really said what it seemed to.

> In spite of...widespread use and exposure, there is little information available on the materials used in fragrance. Fragrance formulas are considered trade secrets...Fragrance is increasingly cited as a trigger in health conditions such as asthma, allergies and migraine headaches. In addition, some fragrance materials have been found to accumulate in adipose tissue and are present in breast milk. Other materials are suspected of being hormone disruptors. The implications are not fully known, as there has been little evaluation of systemic effects...There is little governmental regulation of fragrance. The fragrance industry has in place a system of self-regulation. However, the present system has failed to address many of the emerging concerns.[2]

I felt as if I'd innocently pushed open a door that I hoped would provide the answer to a simple question: why do I get headaches from fragrance? It had instead revealed a dark and dangerous landscape.

~

Fragrance is everywhere. Unless you go out of your way to avoid it, you live in a permanent mist of scented air. Think about the morning routine for a typical woman. She wakes up from eight hours spent in sheets and pyjamas fragranced with detergent. She goes to the bathroom and uses a loo smelling of toilet cleaner. The toilet paper probably smells of flowers, and there might be an air freshener in the bathroom, too. She washes her hands with scented soap. She has a shower using fragranced soap, shampoo and conditioner. She dries herself with a towel radiant with the scent of laundry detergent, and applies fragranced deodorant and moisturiser. She puts on clothes—they smell of the same laundry detergent. She does her hair with fragranced product or hairspray. Each item in her makeup bag—foundation, powder, blusher, lipstick—adds more perfumes. She'll throw the empty packets into a bin with a liner that smells of lemons. To finish, she gives herself a good squirt of her favourite scent.

She hasn't even had breakfast, and already she's been exposed to fragrance in something like fifteen products (sixteen after she's washed up her cereal bowl in scented detergent).

Then she goes out into the world. Say she works in an office. She'll breathe everyone else's choice of perfumes, colognes, deodorants and laundry detergents on the bus. She'll breathe more second-hand scent all day at work.

The air might be freshened there as well, with reed diffusers or a wall-mounted squirter. If she goes shopping at lunchtime she'll get another dose: second-hand fragrance from the other shoppers, plus the room fragrance that many shops now use, and the smell of fragranced candles and soaps in the bed-and-bath shop and the gift shop and the florist. If she shops at a supermarket, she'll get a big hit of detergent smell in the laundry-product aisle. If she takes a taxi, it'll probably have one of the fragrance diffusers. Even after she's gone home for the day she'll still be breathing fragrance, because each one of those sources of second-hand scent will have left a little residue in her hair, on her skin and on her clothes.

We're smelling man-made scents all day, every day. Fragrance is now so pervasive that, as I was finding on that tour, the only way to avoid it is to become—to put it mildly—eccentric. Welcome to Planet Fragrance.

Yet most of us don't know the first thing about fragrance. It had never occurred to me to ask even the most basic questions. What's in those products? Who tests their ingredients? Might there be any problem with breathing them and having them on your skin all the time?

Those questions made me go looking for a book that would give me a few answers. I was sure there'd be one—after all, there are books about longitude and cod and even lead pencils—so I was surprised not to find an accessible, user-friendly source of answers. I found a few books about perfume written by perfume enthusiasts, and a few others written by perfume haters: both kinds were out to persuade me. What I couldn't find was what I wanted: straight-up, reliable information—a book for the general reader that gathered together what people knew about fragrance. I started to see that, if I wanted to read a book like that, I might have to write it first.

The world is full of dodgy information about fragrance, as about most things, so I limited my research to dependable sources. I didn't rely on paraphrases or summaries of the science, but wherever possible I went back to the primary sources: the actual studies, as published in scholarly journals. I found there was no shortage of quality scientific information about fragrance—objective, peer-reviewed and based on properly designed studies—but it's secreted away in specialist publications. I don't pretend to have any deep scientific understanding, but I sought help from science-literate advisors, and found that an interested layperson, with assistance, can generally grasp the basic meaning of what the scientists have found.

Some of the most useful sources were science reviews—where researchers had located all the previous studies about a subject, put them through a rigorous process of assessment, and come up with an overview of the current state of the science. The reviews I used were funded in the interests of public health by the United States, the European

Union and other governments. They were done by panels of independent scientists who've made thorough, objective assessments of the fragrance in everyday products. I consulted the US Food and Drug Administration's website, the reports of the scientific committees of the EU, and the fragrance industry's own statements on their website (ifraorg.org).

As my private research gradually became a public project, I had to make some decisions. One was semantic: what do we mean when we use the word 'fragrance'? That word, or 'perfume' or 'scent', can refer to the stuff in my mother's little bottles, or what you smell when you sniff a rose. It can also refer to the stuff that makes toilet cleaner smell like a pine forest or a lemon. The words are interchangeable, and that blurs an important distinction between the smell of a lemon and the smell of lemon-scented detergent. Unless we're smelling an actual lemon, or the pure essential oil made from it, we're smelling something developed in a lab and made in a factory.

This book is about exactly what that something is.

I'm a novelist, not a science writer. It was my headaches, not any kind of professional urge, that propelled me into research. But what I found in that research has relevance far beyond one novelist who gets headaches. Behind the glamorous public face of fragrance is a private—even secretive—reality that I think is worth sharing.

Fragrance has plenty of friends. The case for it is made every day by people who make money out of it, and people who just love the way it smells.

But there's a downside to fragrance—to do with our health—that you don't hear much about. This book aims to balance things out, not by trying to persuade, but by presenting some of what's known about fragrance. Armed with a bit of information, readers can make up their own minds. Using fragrance is a choice, and my hope is that this book might give people the chance to make that choice an informed one.

CHAPTER 2

HOW MANY OF US ARE OUT THERE?

In 2008 Susan McBride, a senior city planner, was finding life difficult. She worked in an open-plan office for the City of Detroit. Within a few minutes of arriving at work every day, she started to cough and wheeze, and couldn't catch her breath. She consulted doctors, who did tests and confirmed that her symptoms came from being exposed to fragrance.

She discussed the problem with her colleagues. Some of them liked to wear strong perfumes. Others used desktop fragrance diffusers or air fresheners. She explained that any of those smells made it hard for her to breathe. She asked them, as a favour to her, if they'd consider taking their fragrance diffusers home, and waiting till the end of the work day to put on their perfume.

Her colleagues couldn't believe that anyone could get sick from what they thought were lovely scents. Surely she was making a fuss about nothing. Eventually they agreed to remove the fragrance diffusers and air fresheners and little bowls of potpourri. But they wouldn't stop wearing the perfumes they loved.

Susan McBride knew about other offices that had 'low-scent' policies. She went to her supervisor to ask if a policy like that could be brought in for their office, too. She didn't need a completely fragrance-free environment— 'low-scent' would have solved her problem. Her employers refused. They claimed that it would be unconstitutional to stop her colleagues wearing perfume.

One day at work her symptoms got so bad that she collapsed. She was taken to hospital and recovered, but she realised she couldn't return to her workplace while it was full of fragrance. But she wasn't ready to retire. She enjoyed the job, and she needed the money. She got a lawyer with experience in disability claims and took her employer to court.

The Americans with Disabilities Act defines a disability as something that interferes with a 'major life activity'. It's illegal to discriminate against someone on the basis of a disability, and employers are obliged to accommodate an employee's disability as far as they can. The court found that Susan McBride's fragrance allergy was a disability since, as the judge said, her reaction to other people's fragrance 'interfered with the major life activity of breathing'. It found that her employers should have accommodated her disability, and awarded McBride compensation of US$100,000.[1]

Susan McBride's case set a dramatic precedent. Other employers all over North America scrambled to make sure they wouldn't be the next to take a hundred-thousand-dollar hit. Many workplaces in the US and Canada—government offices, private businesses, hospitals, schools and universities among them—now have 'low-scent' or 'fragrance-free' policies.

—

There've been a few similar cases in Australia. In 2014 a public servant, we'll call her Jane Smith, worked for the Department of Human Services. Like Susan McBride, she got sick when she was exposed to the fragrance her colleagues were wearing. She felt nauseous and dizzy, developed a headache, and had difficulty breathing.

Some of her colleagues were sympathetic, and were willing to help her by going easy on the perfume and cologne. But others told her the problem was all in her head. They thought she was just being difficult. How could fragrance make someone sick? Management told her that asking people not to wear fragrance would be discriminatory.

She struggled on until one day, like Susan McBride,

she collapsed at her desk and had to be taken to hospital. It was clear she couldn't go back to work, and she was retired as medically unfit. But, like Susan McBride, she didn't want to retire, and took the employer's insurers to the Administrative Appeals Tribunal for compensation for permanent impairment and non-economic loss.

At the tribunal hearing, five doctors' reports were submitted (one from a professor of immunology) which verified that her symptoms were triggered by fragrance. There was no argument about that. The argument turned on the legal definition of 'impairment'. In the Australian system, claims for permanent impairment can only be paid if the impairment affects at least ten per cent of the whole person. The lawyers for the employer argued that, since her impairment only occurred when she was exposed to fragrance, this was less than ten per cent.

A savvy lawyer for Jane Smith might have mounted an effective challenge against this argument. A lawyer might even have suggested bringing a civil case, based on anti-discrimination laws, rather than going through the AAT and haggling about the arithmetic of impairment. The outcome might well have been different. But Smith couldn't afford a lawyer, and the tribunal found it was 'not satisfied that [the plaintiff's] whole person impairment meets the minimum level of ten per cent in the Act.' She lost her case, and any chance for compensation.[2]

A few similar cases have been successful here. Another

public servant was unable to attend a training seminar because of the participants' fragrance. She won her case and was compensated to the tune of a few thousand dollars.[3] Comparable cases, involving the inability to access shops or other premises because of fragrance, have been settled out of court. But no Australian case has delivered a financial penalty hefty enough to bring about fragrance-free policies in our workplaces, the way Susan McBride's case did in the US.

—

These cases make visible what's usually not: fragrance makes some people seriously sick. Their ability to work or study or even do something as simple as catch a bus is impaired by a choice that someone else is making. But aren't they just an unlucky few? If this problem was widespread, wouldn't we be hearing more about it?

It turns out that Susan McBride and Jane Smith are the tip of an iceberg. Problems caused by fragrance—headaches, breathing difficulties and skin rashes—are common enough that researchers have tried to put a number on them. The numbers are surprising.

—

In the absence of any really clear idea about what causes headaches or how to prevent them, specialist doctors spend a great deal of energy classifying them. There are migraine headaches, tension headaches and cluster headaches. There are stabbing headaches, thunderclap headaches and ice-pick headaches. There are cocaine-induced headaches and caffeine-withdrawal-induced headaches. It seems the only category that's missing is fragrance-induced headaches.

All over the world there are doctors who do nothing but treat headaches. There's an International Headache Society. There are seven headache clinics within a half-hour's drive of my home. There are books and scholarly journals devoted to nothing but headaches. No wonder: the World Health Organization's research indicates that, globally, up to one adult in twenty has a headache every, or nearly every, day.[4]

Not only that: more people are getting more headaches, more often. The prevalence of migraines (especially among younger people) is increasing—one Danish study, comparing rates from 1994 with those from 2002, put this increase at around a third.[5] An American study showed that between 1979 and 1981 there was a 'striking' increase in migraines among people under forty-five. Over those three years, migraines among women increased by more than a third. Among men, the increase was one hundred per cent.[6]

All this suggests that headaches are a problem with no easy answers.

This is awful for people with headaches, but there's

a bright side: it means there's a huge body of research. People unlucky enough to have frequent, severe headaches give specialists a ready-made set of test subjects.

Most headache research is about the particular kind called migraine. These are headaches severe enough to put people to bed in a darkened room for days while powerful drugs attempt to beat back the throbbing pain. Any kind of normal life, including going to work, is out of the question while a migraine is underway. The World Health Organization has put migraine on its list of the ten leading causes of disability, affecting one adult in seven.[7] All studies show that women suffer from migraines much more than men—the prevalence among women is around three times that of men.[8]

Despite decades of research, doctors don't fully understand why migraines happen. But they do know that many people have a migraine trigger. Stress is a common trigger, flickering lights another. Folk wisdom maintains that red wine, chocolate and oranges are migraine triggers, though the research I read didn't mention any of these. But it did point the finger at one: fragrance.

A study from 2001 of patients with migraines found that smells set off headaches in sixty per cent of them. Smells also made headaches worse for sixty-eight per cent of patients. The study concluded that a 'nose–sinus involvement' in headache could be more important than was generally realised.[9] Another study a few years later, of over a thousand migraine sufferers, was more specific about smell. It found that around forty-four per cent of patients named 'perfume or odour' as one of the things that triggered their headaches.[10]

A 2014 study, following on from these findings, was even more specific: of the patients whose migraines were set off by odours, nearly seventy-six per cent named 'perfume' as a trigger. It was the single most common odour trigger for migraines (followed by paint, about forty-two per cent, and petrol, about twenty-eight per cent). The study concluded that 'odorants…especially perfume, may trigger migraine attacks after a few minutes of exposure.'[11]

One more of the many examples was a study that looked at nearly a hundred male migraine sufferers. (Men and women often have different triggers for their migraines. This study was only of men.) Odours were the second most frequent trigger for their migraine attacks (forty-eight per cent), behind stressful situations (fifty-two per cent). Odours were also the second most frequent worsening factor (seventy-three per cent), just behind excessive light (seventy-four per cent). Of the odours, the worst for triggering attacks were perfume, cigarette smoke and fragranced cleaning products.[12]

—

What all these numbers boil down to is that, among people who get migraines, around half get them from fragrance.

But those studies are about the migraine kind of headache. What about the people like me, whose headaches aren't generally bad enough to send us to a doctor, just enough to ruin our day? Putting a number on that question is harder, as you might expect. We aren't neatly corralled in headache clinics with doctors recording our symptoms. Mostly we just pop another headache pill and soldier on.

But a few researchers have come up with some figures. They're not as firm as the headache-clinic numbers, because they're mostly based on people answering questions. Some people, for instance, might think they get a headache from fragrance but really get headaches for another reason, while others might not have put two and two together about their headaches and the smell of fragrance. These surveys could include both over-reporting and under-reporting. Still, the research has come up with some significant numbers.

The New South Wales Department of Health does regular telephone surveys to investigate general health issues. In 2002 one of the questions they asked was: 'Do certain chemical odours or smells regularly make you unwell?' Just under a quarter of the people surveyed said they did.[13] Tantalisingly, the survey didn't specify what kind of chemical odours were causing problems.

But several other surveys did. In 2004 a thousand people in the US were surveyed by random phone interview. Thirty-one per cent reported adverse reactions to fragranced products. A few years later another survey, again of a thousand people, confirmed these results: thirty per cent found scented products worn by other people irritating. Nineteen per cent suffered adverse health effects from air fresheners. Eleven per cent were irritated by scented laundry products.[14]

A third survey, in 2016, found that nearly thirty-five per cent reported health problems such as respiratory difficulties and headaches when exposed to fragrance. (To be exact, eighteen per cent suffered respiratory problems; sixteen per cent had red, watery eyes or nasal congestion; nearly sixteen per cent had headaches; and eight per cent had an asthma attack. Smaller numbers experienced skin problems, cognitive difficulties such as an inability to concentrate, or gastrointestinal problems such as nausea.) Fifteen per cent had lost workdays or a job due to exposure to fragrance in the workplace.[15]

What all that means is startling: out in the general population, fragrance causes health problems for over a third of people. To put that in human terms, if you're one of sixty people on a bus going to the city in the morning, up to twenty people around you may be getting unwell from fragrance.

Many doctors besides my neurologist can confirm this general picture from their experience with patients. Clare Pain's ABC article quoted several of them:

Dr Colin Andrews, a consultant neurologist in

Canberra, with a particular interest in migraines, says perfume is a known migraine trigger for some people.

'Migraine sufferers are very sensitive to strong stimuli of any kind, including perfume. I've had patients who have complained about workmates with strong perfume triggering off their migraines in the workplace'...

Dr Rob Loblay, director of the Allergy Unit at Royal Prince Alfred Hospital in Sydney, says perfumes can be a nasal irritant for people who have hay fever and chronic rhinitis, causing symptoms such as sneezing...'These are the kind of people that don't like walking down the cleaning product aisles in the supermarket,' says Loblay.

'It's quite common...but people often don't complain about it,' he adds...Their symptoms can vary, including nausea, malaise and headaches...

Dr Colin Little, a physician in Melbourne who specialises in allergies, also encounters 'chemically intolerant' patients.

'I see a lot of people who report problems from airborne exposure to perfumes...You have people who in other ways seem well-grounded and in whom there's no reason to believe that they are exaggerating, who report shortness of breath or breathing difficulties or constant sinus pain or things like headache or just anxiety symptoms,' says Little.

He says the biological mechanism behind chemical intolerance to perfume is not yet understood.[16]

Professor Kerryn Phelps, in her popular 'Ask the

Doctor' column in the *Australian Women's Weekly* for November 2016, has this to say about fragranced products such as baby wipes: 'I do see babies who have dermatitis from chemicals in some baby wipes...As a rule, look at the ingredient list on wipes. If it looks like you need a chemistry degree to understand it, use something else. Look for fragrance-free, chemical-free, biodegradable options.'

Susan McBride was rushed to hospital in Detroit because her colleagues' use of fragrance meant she couldn't breathe properly. There's voluminous research, over many years, showing that her problem isn't unusual. Fragrance can trigger coughing, wheezing, tightness in the chest and difficulty breathing. In many cases these symptoms are diagnosed as asthma.

Between ten and fifteen per cent of Australians get asthma. Although it's common, asthma is still not fully understood. It's sometimes triggered by things people are allergic to, like dust mites, but it can also be triggered by irritants, like car exhaust. Exercise can trigger it, and so can infection and stress. Fragrance can trigger it, too.

A study into the link between asthma and odours found that thirty per cent of the respondents with asthma said air fresheners—which emit a lot of fragrance—caused them to have breathing difficulties. Thirty-seven per cent of them found scented products irritating.[17] Another study found that seventeen per cent of the general population experienced breathing difficulties and other health problems when exposed to air fresheners.[18] An earlier research paper noted: 'many patients complain that some odors worsen their asthma. Perfume and cologne are two of the most frequently mentioned offenders.' These researchers found that, out of sixty asthmatic patients, fifty-seven reported respiratory symptoms when they were exposed to common odours, including fragrance.[19]

Back in the 1990s perfume-scented strips were widely used for advertising in magazines. Some people complained that these strips triggered their asthma, and a 1995 study confirmed that they weren't imagining it. Up to a third of the asthmatics in the study experienced symptoms when they were exposed to the scented strips. The researchers found that the more severe the patients' asthma was to begin with, the more likely it was to intensify when they were exposed to fragrance. As they put it: 'Asthmatic exacerbations after perfume challenge occurred in 36%, 17%, and 8% of patients with severe, moderate, and mild asthma, respectively.'[20]

Might these responses be psychosomatic? Several of the early research papers addressed this possibility by pegging the patients' noses, so they couldn't smell the perfume. Even so, the subjects exposed to fragrance experienced asthma, while the ones receiving a placebo didn't.[21]

Two hospital doctors recently published an editorial in the *Canadian Medical Association Journal* in which they drew on the mass of research that's now available on the asthma–fragrance link to plead for fragrance-free hospitals.

> There is emerging evidence that asthma in some cases is primarily aggravated by artificial scents. This is particularly concerning in hospitals, where vulnerable patients with asthma...or skin sensitivities are concentrated. These patients may be involuntarily exposed to artificial scents from staff, other patients and visitors, resulting in worsening of their clinical condition.[22]

Doctors are concerned. So are workplace managers. A survey in the *Journal of Management and Marketing Research* concluded, 'The negative health effects of fragrance to asthmatics are well researched and documented now for many decades,' and urged managers to address the problem.[23]

The American Lung Association is an independent national body representing Americans with respiratory problems. It's in no doubt that fragrance is a common asthma trigger: 'Fragrances from personal care products, air fresheners, candles and cleaning products have been associated with adversely affecting a person's health including headaches, upper respiratory symptoms, shortness of breath, and difficulty with concentration.' Their advice page on creating a 'lung-healthy workplace' includes fragrance on the list of 'indoor air pollutants', and they recommend that a healthy workplace should 'establish a fragrance-free policy for employees and visitors'.[24]

There's been an increase in the prevalence of asthma,

all around the world, that's often referred to in the scientific literature as an 'epidemic'. Patterns in this increase aren't yet understood—there's more asthma in developed countries than in less-developed ones, and females suffer more frequently, and more severely, than males. Researchers are still trying to untangle the causes of the increase and its variation, but they feel that the nature of the increase (the geographic and demographic variations) suggests that environmental and lifestyle factors 'play a large role in the current asthma epidemic'.[25]

One of those environmental and lifestyle factors, of course, is the constant presence of fragrance.

—

Headaches and asthma triggered by fragrance aren't allergic responses, but another kind of symptom is: skin allergies like contact dermatitis and eczema.

With an allergic response, the body meets a substance, recognises it as something that it wants to get rid of, and makes antibodies that help it do so. Once you've made them, they'll swing into action every time you're exposed to that substance. This is why doctors can test for an allergy—they can test for those antibodies.

In a non-allergic response, the body is irritated by the substance, but it doesn't make antibodies, so there's nothing to test for—no objective, measurable biological marker.

(There's a confusing overlap of words with this whole

business. A non-allergic response is often called a 'sensitivity', to distinguish it from a true 'allergy'. But, confusingly, when the body makes antibodies, this is called 'sensitisation'. From the patient's point of view, the big difference is that an allergy can be confirmed objectively, whereas a sensitivity is just a set of symptoms. A better term for a non-allergic reaction is 'intolerance'. In practical terms, though, people who get symptoms from fragrance often find it simpler to call what they have an allergy, as an easily understood shorthand.)

To pin down the exact trigger for a skin allergy, dermatologists do a 'patch test': they make little scratches on the skin and put a drop of various allergy triggers on each patch of scratched skin. If you're allergic to one of the triggers, that scratched patch will get inflamed. One of the drops that dermatologists use is something called Fragrance Mix. There are two standard versions of Fragrance Mix: each contains up to a dozen commonly used fragrance ingredients that are known to cause skin allergies.

Dermatologists think that between one and four per cent of the population has an allergic reaction to fragrance.[26] In Australia, that's between 240,000 and 960,000 individuals. Allergy to fragrance is about four times more common in women than in men, and researchers speculate that this may be because women are generally exposed to more fragranced skin-care and cleaning products than men are.[27]

(One route to sensitisation for men is via their female partners, though, in a process dermatologists wittily call 'consort' dermatitis.)

Skin allergies to fragrance have increased dramatically in the last few decades. They seem a relatively recent phenomenon: they were first indexed in the medical literature in 1957.[28] Since then atopic dermatitis, one common kind of skin allergy, has increased by two to three times in industrialised countries over the past thirty years. It now affects fifteen to thirty per cent of children and up to ten per cent of adults.[29] One allergy study ends by saying, 'the ubiquitous nature of fragrance in modern society, coupled with new and growing markets of fragrance products for children and men, likely contribute[s] to this increase.'[30]

—

Dermatologists seem to enjoy ranking allergens, the way other people rank cricketers or restaurants. In 2006 a group of dermatologists from the North American Contact Dermatitis Group identified the 'top three allergens'—the things that were most likely to cause allergies in their patients. First was nickel sulphate (in things like watches and jewellery), in nineteen per cent of patients. Second was Balsam of Peru, a fragrance ingredient, in nearly twelve per cent. Third was Fragrance Mix, in more than eleven per cent. Since second and third place were both fragrance ingredients, with a combined score of over twenty-three per cent, you could put this differently and say that fragrance was the number-one allergen.[31]

Every year the American Contact Dermatitis Society jocularly names the 'Allergen of the Year'. In 2007 it was fragrance. In 2013 it was methylisothiazolinone, a chemical used as a preservative in fragranced products. In 2014 it was benzophenone, an ultraviolet-light absorber used in fragrance.[32]

If you've got an allergic reaction to fragrance, dermatologists recommend a simple treatment: avoid anything with 'fragrance' or 'parfum' in the list of ingredients. They can't be more specific than that, because most fragranced products contain at least one allergen, and many contain several. A recent survey of fragrance allergy named fifty-four individual chemicals and twenty-eight essential oils as 'established contact allergens in humans'.[33] Even if you can pin down your allergy to just one or two fragrance ingredients, that doesn't help you, because generally they won't be named on the label.

—

The real numbers of people who get sick from fragrance are almost certainly bigger than the studies show. The studies are done, and the numbers counted, in clinics where the symptoms are treated. But for every person who's sick enough to see a specialist, there are many others who just soldier on.

As I was researching this book people would ask me, 'What are you working on now?' At first I was embarrassed to admit what I was doing. Why was a novelist writing a book about fragrance? But as time went on I became braver. 'I get headaches from fragrance, so I'm writing a little book about it,' I'd say.

What happened next was always interesting. Some people would give me the 'uh oh, she's flipped' look. But many people got a certain kind of faraway look on their face. 'Oh, fragrance...' they'd say, and then they'd launch into their fragrance story.

There was the writer friend who couldn't stop sneezing whenever she visited her mother-in-law, because of the fragrance she wore. Another writer chimed in to tell us about the headaches she got from perfumes she described as 'those awful musky ones people wear!' There was the academic who had to sit at the other end of the room when she was in a meeting with a certain colleague who wore a lot of 'product' in his hair. There was the man who'd had to stop taking the lift at work, because the fragranced women in that small space made him start sneezing embarrassingly.

The woman serving me in the fruit shop overheard one of these conversations and cried: 'Oh, that incense they burn everywhere! Makes me sick to the stomach!' There was the ophthalmologist at the Sydney Eye Hospital who gets asthma from fragranced patients. His secretary always warns them beforehand not to wear it but, if they do, the doctor sends them down the hall to the bathroom to wash it off. If that doesn't work, they have to go home and come back another day. The publisher of this book—after urging me to write a novel instead of a book about fragrance—casually mentioned that scented products give him a rash. The editor told me that strong perfumes give him a headache.

What I was hearing seemed like a kind of epidemic. Yet it was a silent one. Some of the people who told me their stories were friends I'd known for years, yet I'd never known this about them. I had the strong impression that fragrance was bothering many people, but no one was talking about it.

It's understandable. Fragrance is supposed to be wonderful. We're supposed to love it. Surely only a weirdo wouldn't enjoy the smell of flowers or pine forests? So everyone puts up with it, thinking they're the only one. Besides, the fragrance conversation is an awkward one to have. How do you tell someone that the way they smell is making you ill?

Yet most people know that they're likely to get a headache or start coughing from the smell of fresh paint. We know there are things in paint that the human body doesn't like, but they're the same things that make the paint stick to the wall, so we put up with the discomfort. We try to minimise it by doing what it says on the tin: make sure there's plenty of ventilation. So it's not such a strange and unfamiliar thing to get sick from certain smells.

With paint, the smell is an unintended by-product. But with fragrance, the smell is the whole point. The things that make that 'good smell' are the same ones that make people sick. That's when you start wondering exactly what's in those bottles. Isn't it just the smell of flowers?

CHAPTER 3

WHAT'S IN
THE BOTTLE?

Until the enlightened later years of the twentieth century, there were no lists of ingredients on things you bought. There were scandals about milk thinned with water, bread bulked out with sawdust. Then laws began to be passed that forced manufacturers to tell people what was in their products. These days, if you want to know what's in your bread or milk you just look at the label. A carton of milk now carries the important information that it contains milk.

I wasn't going to buy perfume, but I went to the fragrance shelves at my local chemist and (holding my breath) took a photo of the back of a run-of-the-mill box of fragrance. At home I enlarged the photo enough to read the list of ingredients: seventeen in all.

The first—the main ingredient—was 'alcohol denat.' A few moments on the internet told me that this was another name for methylated spirits. The second ingredient was 'aqua'—Latin for water. The third was 'parfum (fragrance)'. Well, thank you, but I could have taken a wild guess that a bottle of fragrance would contain fragrance.

Then there were fourteen other ingredients that I'd never heard of: benzyl salicylate, butylphenyl methylpropional, alpha-isomethyl ionone, hydroxycitronellal, amyl cinnamal, citronellol, geraniol, coumarin, linalool, benzyl benzoate, limonene, ethylhexyl methoxycinnamate, diethylamino hydroxybenzoyl hexyl benzoate, M74139.

Several things about this list surprised me. One was that nothing in it said anything about flowers. But what would I know? Maybe alpha-isomethyl ionone is what chemists call roses. Look, darling, I bought you a lovely bunch of alpha-isomethyl ionones!

And I was surprised by how many different elements there were. If what you're trying to get is the smell of flowers, what *are* all those ingredients?

But the thing I really wanted to know about was the fragrance, and the label wasn't going to help me with that beyond giving it a French name. Another few minutes online confirmed what I'd read when I first started looking into all this: the ingredients of fragrance are a trade secret. By law, perfume makers need only put that one word 'parfum' or 'fragrance' on the label (or, just to be sure, both). This list of ingredients wasn't actually a list of ingredients at all.

The International Fragrance Association is the peak organisation for fragrance makers all over the world. It has a website full of information, among which is a list of all the ingredients that go into fragrance—what it calls the 'palette' from which manufacturers pick and choose to make their particular brand.[1] An average 'parfum' might contain a hundred or more of these ingredients. The list is no doubt very interesting and useful, but if you're looking for what's in fragrance it doesn't help you much: there are 2,947 items on it.

Here are the first twenty: Acetanisole; Commiphora erythraea ext.; 4-Methoxybenzoic acid; Grass, hay, ext.; Juniperus virginiana, ext., epoxidised; Octanal dimethyl acetal; 2-Methyl-4-phenyl-2-butyl isobutyrate; 2-Ethylbutyl acetate; Heptanal dimethyl acetal; Hexyl isovalerate; Hexyl 2-methylbutyrate; Styrene; Benzonitrile; Benzyl alcohol; Benzaldehyde; alpha,alpha-Dimethylphenethyl alcohol; alpha,alpha-Dimethylphenethyl butyrate; 3-Hexenyl 2-methylbutanoate; alpha-Methylcinnamaldehyde; Methyl phenylacetate.

A few of these I could take a guess at. Commiphora erythraea, juniper and grass were plants, and styrene sounded familiar—some kind of plastic, was it? But I was out of my depth. If I wanted to know what was in fragrance, I was going to have to start with something simpler.

—

Googling 'fragrance ingredients' unleashes another kind of crowd into your laptop: not the ones who get a headache from fragrance but the ones who adore it. Online I discovered a whole subculture of people who think perfume is somewhere between an art form and a religion. Patrick Süskind, author of the novel *Perfume*, is one of them: his novel is a song of rapture about fragrances. 'Odours have a power of persuasion stronger than that of words, appearance, emotions or will...'[2]

There are hobby parfumiers who buy the separate ingredients and mix their own. There are fragrance consultants who tell you what perfume you should wear. There are people who collect perfumes and compare notes about them online in florid outpourings like this: '[This scent] opens with an orange-plum-grape syrup accord with hints of underlying jasmine...During the late dry-down the composition turns slightly powdery as almond-vanilla-like floral heliotrope takes over control from the vacating jasmine and tuberose, with moderately sweet sandalwood and relatively tame musk adding support through the finish.'

The mystique of the perfume maker and his or her 'nose' is part of the glamour of perfumes: as one commentator puts it, 'mystery is, after all, one of the key intangibles in a perfume's aura.'

Some of these enthusiasts acknowledge that fragrance can give you a headache or make you cough. But many people on these fragrance-fan websites believe that only cheap perfumes might do so. This is because, according to those enthusiasts, they are made with synthetic ingredients, which are petrochemicals. They believe that expensive perfumes are natural, not synthetic, so they would never give anyone a headache.

Sorting out the truth from the tripe on these sites was an interesting journey. It turns out that the 'petrochemical' claim is about as correct as most of the information you find on an internet forum. But it's true that, yes, there are several different kinds of fragrances. Some are natural—scents derived from plants or animals. Others are synthetic—man-made copies of some of the chemicals in the natural fragrances. And there are others that aren't just synthetic versions of natural substances but artificial compounds, new on the planet, with no natural equivalent.

—

Natural smell-chemicals are mostly derived from plants. Over the centuries people have worked out how to chop and squish, soak and boil, evaporate and distil to get a concentrated version of the smell that's generally known as an essential oil. (There are also a few musky ones that come from animals: more on these in a minute.)

I'd always vaguely assumed that an essential oil was just one thing—concentrated smell-of-rose or smell-of-jasmine. What I learned was that an essential oil is a mixture of at least a hundred individual ingredients. Each of these ingredients is a separate chemical.

Rose essential oil, for example, is made up of about a hundred and fifty different chemicals. Some of those chemicals are also in other essential oils—thrifty nature mixes and matches with a few key ingredients rather than reinventing everything for every different plant. What makes rose essential oil smell different from jasmine or lemon are a few ingredients unique to roses, in particular one called beta-damascenone. It's less than one per cent of the whole, but it happens to have a very low 'odour threshold', so the human nose can recognise it straight away as rose.

(For the record, some of the other main compounds in rose essential oil are: pinene, myrcene, 2-phenylethyl alcohol, linalool, cis-rose oxide, nonadecane, farnesol, terpinen-4-ol, geranyl acetate, eugenol, methyl eugenol, citronellol + nerol, geraniol, ionone, docosane and heptadecane.[3])

Natural fragrances—these essential oils—have been used for centuries as perfumes and also as medicines. Aromatherapy and herbalism make great claims for their healing power. A few essential oils have been found to have mild therapeutic effects—jasmine oil may have a slight antidepressant effect, for instance.[4] But many of the claims made for essential oils aren't backed up by much research.

What is known, though, is that some essential oils

are definitely *not* good for you. Pennyroyal oil and sassa-fras oil have been shown to be carcinogenic.[5] Balsam of Peru is a powerful sensitiser, causing skin allergies.[6] A study from 2007 suggests that lavender oil can have undesirable hormonal effects.[7] Some doubt has been cast on this research, but a major American cancer-treatment centre nevertheless says that long-term use of lavender oil 'should be avoided in patients with hormonal cancers.'[8] Even aromatherapists warn against using certain plant extracts, and the International Fragrance Association lists about twenty essential oils that are either prohibited or restricted for use in fragrance.[9]

So—contrary to common belief—not all essential oils are good for us. When you think about why a plant makes these substances you can start to understand the reasons. Smells are a matter of life and death for something stuck in the ground, unable either to chase or flee. Nice smells are one of a plant's few ways of attracting insects that will fertilise it. Without them, the plant can't reproduce. Nasty smells and tastes are a good way for plants to protect themselves and their precious seeds. Some plant compounds kill fungus and bacteria. Others protect the plant by tasting bitter or being poisonous.

Animals make smell-chemicals for the same reasons. Skunk smell turns away attackers as effectively as tear gas. At the other extreme, the smell given out by the male deer is a big turn-on for female deer. (Clue to what kind of turn-on: the word 'musk' comes from the Sanskrit word for 'testicle'.) If you're a boy deer, the more pungent the musk smell you can make, the more offspring you're likely to have.

The bottom line is that natural fragrances aren't just a generous gift from nature to make our lives nicer. They're powerful and sometimes harmful substances. But humans have evolved alongside plants and animals. We know to steer clear of bad smells and bitter tastes, and, in the amounts found in nature, the nice smell-substances don't have any harmful effects on us. After all, we can smell a rose without getting a headache.

The trouble is that over the millennia we've learned to concentrate those already potent substances. It takes up to four thousand kilos of rose petals to make a kilo of rose essential oil.[10] That means when we smell rose essential oil, we're getting thousands of times more of all those rose chemicals than when we smell an actual rose.

Essential oils are natural, in the sense that they're made from plants and animals. But they're also unnatural, in the sense that smelling them is like having the scent of hundreds of roses up your nose rather than just the scent of one.

—

Archaeologists have found perfume bottles dating back four thousand years. Between then and about 1887, anyone who wanted to make nice-smelling stuff had to work with plants and deer testicles. Then chemistry suddenly took a giant leap forward and clever people learned how to make synthetic fragrances.

First they worked out how to separate the natural fragrances into their different ingredients: to isolate the beta-damascenone from the other 149 things in rose oil, for instance. Once they'd got an individual ingredient out on its own, they could discover exactly what it was made of. They were able to establish that every molecule of beta-damascenone is a combination of thirteen atoms of carbon, eighteen atoms of hydrogen and one atom of oxygen, all stuck together in a particular way. The shorthand is the chemical formula $C_{13}H_{18}O$. The longhand is (E)-1-(2,6,6-Trimethyl-1-cyclohexa-1,3-dienyl)but-2-en-1-one. (That tells a chemist not just what's in it but exactly how the bits are stuck together.) In chemistry speak, that mouthful is the name of the rose.

Getting nature to show them exactly what she was made of was interesting, but it wasn't going to make anyone rich. What the chemists were really hoping to do was bypass her altogether. Essential oils are expensive: five millilitres of natural rose oil—about a teaspoonful—costs around $300. The holy grail of perfume chemistry was to create the smell of a rose from something that was cheaper than roses.

The theory of that was straightforward. Wherever carbon occurs in the world—whatever mix it's part of—the carbon atoms are always exactly the same stuff. The same goes for hydrogen and oxygen. So when chemists want to get thirteen atoms of carbon to start making beta-damascenone, they can get them from anything that contains carbon, even if the carbon is mixed up with other things. They can get the hydrogen from anything that contains hydrogen, and the same goes for the oxygen.

Hydrocarbons are compounds that, as the name implies, contain hydrogen and carbon. Petrochemicals are one cheap and plentiful kind of hydrocarbon. The chemists worked out how to get at the carbon and the hydrogen in the hydrocarbons, and got the oxygen out of plain old air. Cooking up the three ingredients to make them stick together in just the right way was trickier, but they worked it out. The result was (E)-1-(2,6,6-Trimethyl-1-cyclohexa-1,3-dienyl)but-2-en-1-one: rose smell.

Once you get the exact number of atoms of carbon, hydrogen and oxygen to line up in exactly the right way, it doesn't matter what they were part of before. The beta-damascenone made by a rose is exactly the same stuff as the beta-damascenone that they make in the lab. As some labels say, it's 'nature-identical'. So, yes, synthetic fragrances might be *derived from* petrochemicals, but that doesn't mean they *are* petrochemicals.

The difference is that for $200 you get a kilo of synthetic beta-damascenone, while for the same money you get less than a teaspoonful of rose essential oil—and the beta-damascenone is only one per cent of that teaspoonful.

Perfumes made before the Second World War usually contained a fair proportion of natural ingredients, because many of the synthetics hadn't yet been developed. But these days fragrance makers can produce rivers of fragrance for the cost of a single basket of hand-picked rose petals. The manufacturers are able to produce their commodity on a gigantic scale, rather than work at the cottage-industry level of the past, and reap the benefits of mass production.

This imposes a certain financial logic on the manufacturers, who, behind all the mystique about 'noses', are sensible business people. Whether they're making this year's top fashion perfume or a pine-smelling toilet cleaner, these days all fragrance makers use synthetics.

The formulation of the expensive perfume is more refined than the formulation of the toilet cleaner, of course. The mix of ingredients is more subtle and pleasing to the nose. And some expensive perfumes may still contain a few natural ingredients. But, thanks to the labelling laws, no one but the manufacturer will ever know. And—not to be cynical—why would perfume makers use essential oils when the synthetic version is a hundred or a thousand times cheaper, and the recipe is a trade secret?

The unromantic fact is that today's fragrances are made from things with names like acetaldehyde ethyl cis-3-hexenyl acetal. Write a novel about *that*, I dare you!

—

So an individual synthetic fragrance chemical is identical to the same chemical found in a plant or animal. But somewhere in the persistent myths about synthetics versus naturals is a tiny grain of truth. It's got nothing to do with petrochemicals, though. It's about concentration and combination.

In nature you can never get a really big hit of beta-damascenone. No matter how intensely you concentrate rose juice, the beta-damascenone is only ever going to be about one per cent of it. But once you've got a test tube full of man-made beta-damascenone, you can use as much of it as you like in the fragrance you're making. You might decide that a weedy one per cent concentration isn't enough. You might decide that ten or twenty or fifty per cent will make a better product, with a larger market.

There may be nothing wrong with inhaling beta-damascenone at ten or twenty or fifty times the natural proportion. But, as we've seen, these smell-chemicals are potent things. Something that's harmless at one per cent may not be so harmless at a higher concentration.

There's another big difference between the synthetic version of beta-damascenone and the essential oil that contains it. In the essential oil, beta-damascenone is one of about a hundred and fifty ingredients. In a synthetic fragrance, the manufacturers may not have added any of those 149 other compounds, so the beta-damascenone will be without its usual companion chemicals.

Scientists are still investigating some of those micro-ingredients and finding out how they work together. They know that some of them buffer the harmful effects of others (by acting as an antioxidant, for example). But the jobs and potential protective effects of some of the others are still a mystery. Leaving them out might be robbing us of a protection that the natural mixture provides.

Maybe there's no problem with inhaling high concentrations of beta-damascenone (or any other smell-chemical), and it might be fine to smell it without its natural companion chemicals. But it's a new thing in the world. Humans evolved alongside plants, so the smell of a rose—that mix of a hundred and fifty chemicals, with beta-damascenone just one per cent of the mix—is something homo sapiens has been living with for hundreds of thousands of years. Manipulating the concentration and the mix in the way we can now do might be good, it might be bad, or it might make no difference to anything. But it's certainly in the nature of an experiment, with humans as the lab rats.

—

A funny thing happened with some of those synthetic fragrances: sometimes they behaved weirdly once they were isolated. Deer musk is a case in point. Chemists worked out how to make a synthetic version of the main smell-chemical, something called muscone. Much cheaper, and no outrage from animal lovers about the deer you had to kill to get the real thing. But deer musk smells good, while muscone apparently has what is sometimes delicately described as a faecal or animalic aroma. In the natural substance, the muscone is only one of many ingredients, and it's likely that they all work together in some way to cancel out that undesirable whiff. The challenge was to get rid of it in the synthetic version.

The chemists thought it might be a particular bit of the muscone molecule—the 'nitro function'—that was causing the faecal effect. To get rid of it, they set about tinkering with the molecule, changing its structure so that it was no longer the same as the one in the natural substance. Once you've tinkered with the actual shape of a molecule, you've invented something that doesn't exist in nature. You've produced a completely new fragrance chemical—an artificial substance with no natural equivalent.

Here's how a distinguished contemporary fragrance chemist, Philip Kraft, describes the kind of thing they do. 'You can play with conformational elements that drive a certain molecular shape—for example you can introduce structural elements that cause a molecule to fold itself into a horseshoe shape and thus smell musky.' Adding rigidity to molecules often gives more defined odour notes, he adds. 'Or you can make the molecule more flexible to add new by-notes—you can cut some parts to make it lighter and more diffusive while conserving its overall shape.'[11]

These new chemicals are patented by the company that developed them and are given brand names such as Tonalide and Galaxolide.

The first artificial musks were invented around the 1920s. They must have seemed like a gift from the gods, because the core ingredients of soap and detergent don't smell very good. All soaps are based on some kind of oil or fat. The cheapest kinds are some version of mineral oil (a petroleum product) or animal fat (tallow). Things made from those raw ingredients aren't going to smell like clean laundry. They're going to retain some of the smell of the oil or fat they've been made from. Musks are there to hide that smell.

From about the 1950s, artificial musks started to be used in huge quantities. Since then, chemists have continued to tweak and tinker, and come up with more and more varieties of them. Musks have a pleasant aroma often described as 'marine' or 'clean laundry'. These days artificial musks are in all mainstream laundry detergents— they're what makes the laundry aisle of the supermarket smell the way it does. We now expect laundry to have that 'clean laundry' smell, not realising that it's not the actual smell of clean laundry, but the smell of a man-made chemical that smells a bit like clean laundry.

—

When manufacturers set out to make a fragrance, they'll combine anything up to several hundred different smell ingredients. That smell will change over time, as the separate ingredients evaporate and combine with air. One of the challenges of the parfumier is to control this change. The classic way to go about it is to put a smell together that has 'top notes', 'middle notes' and 'base notes'.

The musical metaphor romanticises a plain physical fact: different chemicals evaporate at different rates. The top notes in a fragrance are the ingredients that evaporate most quickly. When you first put a perfume on your skin, the top notes will be the most evident because they're evaporating (becoming a vapour in the air) straight away. Middle notes vaporise more slowly, and base notes vaporise more slowly still. When the fragrance has been on your skin for some time, you'll mostly be smelling the slow-to-evaporate ingredients. A fragrance that 'lasts longer' will tend to have plenty of base notes to go on filling the air after the top notes have evaporated.

The quality of 'travelling' or 'projecting' is highly valued in a perfume—it seems people *want* everyone else to smell them coming. (The boast on one retail website is that perfume will 'provide women with the ability to draw a whole new level of attention to themselves'.) To make a fragrance travel, you make sure it has top notes that vaporise especially easily. Warm skin sets the molecules leaping out into the air, spreading away from the source.

The balance and timing of all this is part of the parfumier's skill, but it's not magic. It's just the physics and chemistry of the ingredients.

—

What else goes into a fragrance? Apart from the way it smells, there are other more mundane considerations. Some of the fragrance chemicals are sensitive to ultraviolet light, causing the smell to change from rich to rancid, so perfume makers often add a chemical that blocks ultraviolet rays. Other fragrance chemicals react with the oxygen in the air and turn into bad smells that way, so an antioxidant is added, too. The colour of a fragrance can change from a lovely clear yellow to a turbid khaki, so a colour stabiliser might be needed. Or the colour mightn't be attractive in the first place, so a colouring needs to be added to make it look pretty.

Then something's needed to mix all these things together and keep them mixed: a solvent and preservative. Originally the best thing for this was ambergris, a strange fatty substance that whales produce. Ambergris was eye-wateringly expensive, though, and isn't used any more because a much cheaper thing has been invented that does the same job: a chemical called diethyl phthalate. If a product has 'fragrance' on the label, it almost certainly contains diethyl phthalate, otherwise known as DEP.

So, in a general way, that's what's in the bottle: something that smells nice, a few things to stop the nice smell going off, and something to blend everything together.

For the last four thousand years or so, humans have gone to a lot of trouble to make nice smells. I was starting to wonder just why. What is it about the sense of smell that makes it all worthwhile?

CHAPTER 4

WHAT NOSES KNOW

Years ago I read a novel that I remember nothing about except one scene, in which a nerdy little boy at his class 'show and tell' presents what he thinks is an interesting fact. He tells his classmates that when you smell something, it's because actual bits of it are getting up your nose. He mentions going into the bathroom after his father's been in there. Cries of disgust from the other kids.

When I read this I thought the writer (whose name I've forgotten, along with everything else about the book) was making up that interesting fact. But he was right. When you smell something, it's because little bits of it have just gone up your nose.

What happens up there in the far reaches of your upper nose is that those bits—molecules of smell-chemicals—lock on to the ends of stringy nerve receptors. These are connected to the brain through a grid of bone at the top of the nose. (I picture a tea strainer with bits of dental floss hanging down through the holes.) Those nerve receptors are specially made to recognise smell molecules, and when they do they send a signal up through the grid. Just on the other side of that grid is the brain's olfactory bulb. Its job is to gather those smell signals and send them to other parts of the brain to be interpreted. Ah, this is the smell of those little cakes I used to eat when I was a child. Oh, Dad's been in the bathroom.

All that stands between the brain and the outside world is that tea strainer. This is the only place in the body where the brain is directly connected to the outside world. So why has nature left this chink in its armour? Because smelling things fast used to be something that could save your life. In the jungle, where you can't see much, you might have smelled the tiger before you saw him. This is why reaction to a smell—good or bad—is so rapid. There's no mucking around with getting the molecules into your bloodstream. The smell-chemical signals go straight to your brain. When there's a tiger around, sensitivity as well as speed matters: sniffing tiny amounts of tiger aroma could be a lifesaver. Our noses are so sensitive that they only need a few molecules to know what's out there.

What this means is that it's not the *smell* of fragrance

that's the problem—it's the actual *molecules* of the fragrance chemicals that are going into your nose. Even people who've lost their sense of smell—people with anosmia—can still be physically affected by fragrance.

In these largely tiger-free days, our sense of smell might seem something of a biological luxury. But it evolved as part of our system of distinguishing between things that are good for our survival and things that aren't, and it still works like that. It's the sniff test that warns us not to eat an oyster that's been in the fridge a day too long. It's our nose that tells us to open the window when we're painting the room.

It's no accident that people whose health is fragile can become hypersensitive to smells. Hospitals discourage strongly scented flowers because sick people often don't like them. People going through chemotherapy generally want to avoid strong smells. And pregnant women are famous for becoming crazily sensitive to smells. (One woman told me she always knew when she was pregnant, even before the test results came through, because she suddenly couldn't stand the smell of other people's perfume.)

—

Nature wants our sense of smell to stay alert, and it's come up with a clever little refinement. You know how you walk into a house where there are sixteen cats, and the smell just about knocks you down? You wonder how the people who live there can stand it, and then you realise they can't smell it. Ten minutes later, neither can you—your nose has stopped sending pay-attention messages to the brain. This is olfactory fatigue.

It's another leftover survival mechanism. If something's been around for a few minutes and it hasn't hurt you, chances are it's okay. But you need to be able to detect something new in the environment, just in case it's a bad kind of new. So the brain turns the volume down on the proved-to-be-safe smell, so it can be sure of picking up the maybe-dangerous smell.

This is why, the more fragrance you're exposed to, the less you smell it. Most people aren't aware of the smell of fragrance in their freshly washed laundry or their shampoo. And a few minutes after they've put their perfume on, they can't smell it any more. 'Oh, it's faded,' they think, and give themselves another squirt. But it's not the smell that's faded, it's just that their smell receptors are tuning it out. If you use perfume or fragranced products, there's a kind of bracket-creep effect: you have to use more and more so you can continue to smell it.

Learning about all this, I could find it in myself to forgive that woman behind me at *Don Giovanni*. Her problem wasn't necessarily that she was selfishly unaware: it was just that she was experiencing olfactory fatigue. For her, the scent she'd just squirted on was nothing but a faint pleasant whiff. It was the people around her who found it overpowering.

There's one other survival-of-the-species thing going on with the sense of smell. It's got to do with sex—or, at least, with reproduction.

Every human makes something called Human Leukocyte Antigen. It's an important part of our immune system. There are a few different variants of HLA, and scientists have found that two people with different HLA will tend to have children with stronger immune systems than people who share the same kind. And it turns out that, at a subliminal level, a woman can *smell* whether a man has different HLA from her. She won't be aware of it consciously, but her nose will have quietly informed her brain that this one would be good mating material. What's more, her nose is sending this information just when it's most useful. A woman's sense of smell is sharpest around the time of the month when she's most fertile.[1]

Something equally useful happens for men. Their olfactory bulbs can pick up when a woman is fertile. In survival-of-the-fittest terms, it makes sense to wait for the right moment to give sperm its best shot at connecting with a ripe egg.[2]

Perfume ads like to suggest that they've discovered the pheromone that will turn on the opposite sex. You can't blame them for trying, but we don't need to buy a 'signature scent'. All this research shows that we've already got one.

One thing that's bad for survival of the species is incest, and the nose has a role in warning about that, too. It's hard to believe, but apparently we can unconsciously recognise—by smell—someone who's our brother, sister, mother, father or child.[3]

Once sperm and egg have found promising partners and produced offspring, the next trick is to keep that offspring alive. Once again the nose helps out. Baby lambs die if their mothers can't pick them out of a crowd by smell—they do all look the same, after all—and this works for humans, too, though we may not consciously realise it. In one study, '90% of women tested…identified their newborns by olfactory cues after only 10 min–1 hour exposure to their infants. All of the women tested recognized their babies' odor after exposure periods greater than 1 hour.'[4]

The babies can recognise their mothers by smell, too. Not surprisingly, this very basic sense of recognition and knowing where you belong is thought to be important for mother–baby bonding. Researchers put it this way: 'naturally occurring odors play an important role in mediating infant behavior…Olfactory recognition may be implicated in the early stages of the mother–infant attachment process, when the newborns learn to recognize the[ir] own mother's unique odor signature.'[5] Other researchers studied newborns experiencing pain and found that, if the babies were exposed to the odours from their own mother's milk, their distress was less than if they were exposed to the odours of the milk of another woman.[6]

People who deal every day with newborns know how important it is that smell has a chance to do its baby–parent bonding work. A young friend of mine recently had a baby at one of the Canberra hospitals. The nurses told her and her partner not to use anything fragranced for a few weeks before the birth, and for a few months afterwards. No perfume and no colognes, only fragrance-free deodorant and shampoo, and unscented laundry powder. The way the nurses put it was: 'Bub needs to learn the smell of Mum and Dad.'

It made me wonder what happens if Bub doesn't learn those smells. If all you can smell is fragrance out of a bottle, how do you learn to recognise Mum and Dad, and enjoy the security and reassurance that familiarity provides? It's not something scientists can readily quantify. But it's worth asking: if the smell of Mum and Dad is drowned out by the smell of artificial fragrance, what happens to Bub's sense of belonging?

Which took me back to the big question: exactly what is in all those fragrances?

CHAPTER 5

BEHIND THE LABEL

In the normal way of things, no one knows what's in a particular fragrance except the people who make it. But luckily scientists have machines that can analyse what's in a substance. Many researchers have put fragranced products through a process called matrix solid-phase dispersion, followed by gas chromatography–mass spectrometry. This lets them get past the locked door of the 'parfum' word on the label to see what's in the bottle.

Anne Steinemann, a professor of civil engineering at the University of Melbourne, wanted to unlock that door because she was researching 'sick building syndrome'. From the 1970s on, buildings—especially office blocks— became more airtight to save on heating and cooling costs. That was good, but there was an unintended, mysterious consequence: the people who worked in those buildings sometimes became unwell. It wasn't the buildings that were sick: it was the people. Some of them developed severe symptoms like migraines and asthma, but many more complained of being always tired and generally unwell. They found it hard to concentrate and make decisions.

This was having a bad effect on productivity, so scientists and engineers were called in to investigate. They found those airtight buildings were accumulating substances called volatile organic compounds, or VOCs, which they thought might be connected with the symptoms people were complaining of. (VOCs are called volatile because they evaporate easily and float around in the air. They're called organic not because they're clean and green, but because chemists use the word to mean that the compound contains carbon.)

Hundreds of things in an office give out VOCs, in particular all the man-made items: carpet, glue, photocopiers, paint and plastics. It took a while to realise that fragrance was another important source of VOCs, evaporating out of all the scented personal products people used on themselves, as well as the cleaning products, air fresheners, potpourri and fragrance diffusers. Now that the buildings were airtight, these VOCs were endlessly recycling through the closed loop of the air-conditioning system, rather than being diluted—as in the past—by fresh air from open windows.

Other research had suggested that the use of fragranced products in buildings could result in concentrations of VOCs that exceeded government guidelines designed to protect human health.[1] Professor Steinemann wanted to find out exactly which VOCs might be in those products, and at what concentrations, to investigate whether they might be part of what was making office workers sick.

In 2009 she and her colleagues—she was then at the University of Washington—took a representative sample of twenty-five everyday fragranced products and gave them the gas-chromatography treatment. They were only looking for the volatile fragrances—there are others that are 'semi-volatile organic compounds', which include artificial musks, but their analysis didn't cover them. They also didn't sample 'fine fragrances', that is, perfume. They found 133 different VOCs coming out of the products. Each product contained between six and twenty VOCs. The most common (in ninety-two per cent of the products) was limonene. The second most common (eighty-four per cent) was alpha-pinene and the third most common (eighty per cent) was beta-pinene.[2]

These three chemicals, of similar molecular structure, all belong to a group called terpenes. Terpenes often smell nice and in themselves they generally don't cause serious health problems, though they can irritate eyes and lungs. But they have one very unfortunate characteristic—when they come in contact with air, they go through a chemical reaction that results in formaldehyde. Formaldehyde is classed as a 'known human carcinogen'.[3] Another carcinogen, acetaldehyde, was found in about a third of the products. Several products also emitted 1,4-dioxane and methylene chloride. These are also carcinogens.[4]

As well as these carcinogens, many of the most commonly occurring chemicals identified in the study produce other less serious health effects: redness, painful and burning eyes, dry and irritated skin, skin allergies, cough, headache, and fatigue.[5] Of the 133 VOCs found in these products, twenty-four are classified as toxic or hazardous by American regulators. Every product that was analysed contained at least one of these chemicals.[6]

No wonder all those office workers were wilting at their desks.

—

Five years later Professor Steinemann did another, similar analysis, of a different sample of fragranced products. This time she included products that advertised themselves as 'green'—they had words like 'eco', 'organic' or 'natural' on the label—and some that were called 'fragrance-free'.

The exact line-up of the VOCs was different in the second study, as the sample was of different products, but the sum total was just as bad. Between them, the thirty-seven products emitted 156 different VOCs. Forty-two of these are classified as toxic or hazardous by American regulators. Each product emitted at least one of these chemicals. About half the products emitted at least one carcinogen: formaldehyde, 1,4-dioxane or methylene chloride.[7]

Steinemann found there was no significant difference between the VOCs from regular products and 'green' ones. The 'fragrance-free' ones, though, had a big advantage over the regular ones—none of them contained the formaldehyde-producing terpenes.

The engineering solution to 'sick building syndrome' was to make sure the air-conditioning system took in enough fresh air to dilute the VOCs. Codes were introduced that tried to balance the efficiency of buildings against the needs of the humans inside them.

Employers and owners of office buildings had the resources to locate the problem and fix it. It was worth their while because brain-fogged workers were bad for business. The analysis was done, the culprits found, and the action was taken. But for those of us whose households are sadly lacking in machines that perform matrix solid-phase dispersion followed by gas chromatography–mass spectrometry, it's not so simple. We can't know what's in our face cream or air freshener, so we can't make an informed decision about whether to go on using it. Unlike those office workers, we continue to be exposed to whatever the manufacturers have chosen to include.

—

Anne Steinemann's results made me curious about all the things listed on the label of the perfume I'd photographed at the chemist's. Of course, there were the mystery ingredients that lay behind the word 'parfum', but there were sixteen other ingredients identified by name.

One was rather expensive water. That left fifteen other ingredients. Six of them, I discovered, are skin, eye and respiratory tract irritants: the denatured alcohol (otherwise known as methylated spirits),[8] plus butylphenyl methylpropional,[9] hydroxycitronellal,[10] geraniol,[11] benzyl benzoate[12] and limonene.

Ten of the fifteen are on a list of chemicals that are restricted by European Commission consumer-protection authorities. They're on that list because the health problems they cause (mostly skin allergies) are established beyond argument. There are twenty-six chemicals on the list, which means more than a third of those restricted chemicals are in this bottle. (The restriction doesn't stop them being used in consumer products. It just means they have to be declared by name on the label and used in specified concentrations.) The ten restricted ingredients in this fragrance are: benzyl salicylate, alpha-isomethyl ionone, hydroxycitronellal, amyl cinnamal, citronellol, geraniol, coumarin, linalool, benzyl benzoate and limonene.[13]

Nine (not always the same ones) are on a list of chemicals that are restricted by the manufacturers of fragrance. Those nine chemicals are on that list because they can cause bad health effects—again, mostly skin allergies. But, as with the European Commission's restricted list, the fragrance industry's restriction doesn't stop the chemicals being used in fragranced products—it just limits the concentration. The idea behind this is that if there's only a small amount of them in the bottle, they won't do any harm. These restricted ingredients in this bottle are: benzyl salicylate, butylphenyl methylpropional, alpha-isomethyl ionone, hydroxycitronellal, amyl cinnamal, citronellol, geraniol, coumarin and benzyl benzoate.[14]

Three ingredients in this perfume are in Fragrance Mix I, used by dermatologists to test for skin allergy. This means they're well known for producing contact dermatitis or eczema. They are hydroxycitronellal, amyl cinnamal and geraniol.[15]

Three chemicals named on the label are hormone disruptors: something about their structure means the body thinks they are the same as the oestrogen it makes itself. Too much or too little oestrogen can do strange and unwanted things in the bodies of both men and women. The hormone disruptors in this bottle are ethylhexyl methoxycinnamate,[16] benzyl salicylate and benzyl benzoate.[17]

One ingredient, M74139, is a mystery—I wasn't able to find anything about it. It may be a colouring.

What all this means is that, of the sixteen named ingredients in this scent, only three aren't well-recognised health risks. These are the ultraviolet-light blocker diethylamino hydroxybenzoyl hexyl benzoate,[18] the mystery ingredient and good old aqua.

—

By the time I'd finished researching the things in that little bottle of perfume, my inner sceptic was starting to be activated, just as yours may be. People are buying and using this stuff every day. But there are no warnings on the package. Had I somehow got the wrong end of the stick? If fragrance is really this bad for us, how come you can buy it in any supermarket?

CHAPTER 6

WHO'S TESTING FRAGRANCE?

I'd always assumed that anything out there on the market has been tested for safety. Gone are the days when Coca-Cola contained cocaine and Radithor (a tonic supposed to cure impotence) contained enough radium to kill you.

These days, consumers are protected by governments. In Australia, safety is looked after by the Australian Competition and Consumer Commission. The ACCC classifies fragrance as a cosmetic and has issued regulations about it. The ingredients in fragrance must be 'prominently shown and clearly legible', so that 'a person with normal vision can read the whole ingredients list on the product or container without error, strain or difficulty.'[1] Hence all that clearly legible writing on the back of your shampoo bottle.

Three cheers for that law! But there's no law without its loophole, and the loophole with ingredient-labelling laws is the concept of the protection of trade secrets. This exempts fragrance makers from telling consumers what's in the 'parfum' part of their product.

The reasoning behind trade-secret protection is that manufacturers—of all sorts of things, not just fragrance—spend a lot of money on research, in order to make something unique. If anyone could come along and copy it, their business would suffer. (Actually, with fragrances this does happen: perfume makers use those gas-chromatography machines to tell them what's in a competitor's product. That's why we can buy cheap knock-offs of expensive scents. The protection of trade secrets won't necessarily stop someone in the fragrance business copying a popular scent, but it does stop consumers knowing what's in it.)

In any case, labels are as far as the ACCC goes. What I wanted to know was: does anyone test fragrance for safety?

—

Another federal-government body, the National Industrial Chemicals Notification and Assessment Scheme, has the job of checking that the things we buy are safe. Naive inhabitant of the nanny state that I am, straight away I formed a vivid and reassuring picture of a lab somewhere in Canberra full of white-coated boffins peering at test tubes.

A little research taught me that I'd have to revise that picture. No white coats, no test tubes—NICNAS doesn't have the resources to do any actual testing. What it does is to keep a 'register' of the industrial chemicals used in Australia. There are about forty thousand chemicals on the register. It also 'assesses' some of the chemicals on that register. This means that it looks at the results of tests that have already been done on the chemical in question. On the basis of that information, it decides how safe or unsafe the chemical is. Of the forty thousand chemicals on the register, NICNAS has assessed about three thousand.[2]

The way it does this assessment is by gathering together all the available information about the chemical. This information comes from various sources and it's a bit of a mixed bag. One source is the people who make the chemicals—they're obliged to issue a Safety Data Sheet that contains safety information. But Safety Data Sheets don't always offer full disclosure. Another source is the International Chemical Safety Cards. These are issued by the World Health Organization—but they're designed to protect workers exposed to large amounts of the stuff, rather than consumers.

The European Commission's Scientific Committee on Consumer Safety tests some fragrance ingredients and publishes its results. So does the US National Toxicology Program. Independent scientists in universities test some of them, too, if they can get funding from somewhere.

And one of the main sources of information about fragrance chemicals are the tests funded by the International Fragrance Association—that is, tests done by the same people who make and sell the products. For a good number of fragrance chemicals, these tests appear to be the *only* ones that have been done, and therefore virtually the only source of information NICNAS has about them when it assesses a chemical.

Even worse, for quite a few chemicals there is simply no data of any kind. With the best will in the world, NICNAS can't assess them for safety. All it can do is to signal that this gap exists. It does this by publishing a list of nineteen 'data-poor fragrance chemicals' on its website. NICNAS can't say whether or not these chemicals are safe. Their possible effects on living things, including humans, is unknown. No one even knows what quantities of these chemicals are imported into Australia, which could give us some idea as to our exposure to them. And, of course, no one but the fragrance manufacturers knows which products they're in.[3]

Hemmed in by these limitations, NICNAS looks at the sources of information about a chemical and makes the best ruling it can about the way it should be used. This might mean a safety label like the one telling you not to let the shampoo get in your eyes. Or it might mean a restriction about how much of the chemical can be in a product. A few chemicals have been banned, but that seems to be a last resort, even for chemicals with well-established dangers.

An example is formaldehyde, which is emitted by many cosmetics. NICNAS has assessed formaldehyde. According to a research survey by the ACCC that cites the assessment, 'Exposure to formaldehyde through the use of cosmetics is a recognised hazard...the critical health effects of formaldehyde for risk characterisation are sensory irritation, skin sensitisation and carcinogenicity.'[4] The US National Toxicology Program includes formaldehyde on its list of chemicals 'Known to be Human Carcinogens'.[5]

But NICNAS hasn't banned formaldehyde (nor have the American regulators). It's arrived at a concentration it considers to be safe in cosmetics: 'Most cosmetics may not contain more than 0.05 per cent free formaldehyde unless they are labelled with the warning statement: contains formaldehyde.' If the product label contains those words, the product can contain up to 0.2 per cent (up to five per cent in nail hardeners).[6] A consumer who happens to know that formaldehyde is a health risk may read the label carefully and choose not to use the product. Less-informed consumers mightn't know whether containing formaldehyde is a good thing or a bad thing.

—

There's a form on the NICNAS website for people who want to apply to have a chemical assessed. If I believed that something in fragrance was giving me headaches, and I wanted NICNAS to investigate, in theory I'd be free to apply. But the first hurdle would be that word 'fragrance' or 'parfum' on the bottle. Before I could get to first base and name a chemical for NICNAS to assess, I'd have to take the fragrance to a chemist and get them to run it through their gas-chromatography machine. That would identify not just the ingredients named on the label, but also the unnamed ones in the 'parfum'. There could be a hundred or more chemicals. Then I'd have to test each one of them separately on myself to find which one was giving me the headache. (Ideally I'd also test combinations of them, in case the individual chemicals were okay but caused problems when they interacted.)

Once I'd fingered a chemical, NICNAS might assess it, but their assessment would be based at least in part on information provided by the people who make or sell products containing the chemical. Once NICNAS had made its ruling, those people could appeal the assessment and bring forward more data in its defence. The manufacturers might possibly have more resources behind them than a novelist with a headache.

—

In the US, the government's National Toxicology Program tests some fragrance ingredients. It publishes its findings and maintains online lists of dangerous substances. But another body has the job of deciding how to act on that science. That regulator—the Food and Drug Administration—has limited power. The FDA website spells it out:

> The law does not require cosmetic products and ingredients, other than color additives, to have FDA approval before they go on the market...Neither the law nor FDA regulations require specific tests to demonstrate the safety of individual products or ingredients. The law also does not require cosmetic companies to share their safety information with FDA...the burden is on FDA to prove that a particular product or ingredient is harmful when used as intended.[7]

So a chemical goes on the market without having been tested. If there's a suspicion that it might be bad for people, the manufacturer doesn't have to prove it's safe: the FDA has to prove its risks. Even after that it has to jump several hurdles before it can act to regulate a chemical. How 'harmful' does a chemical have to be before it's regulated? What does 'used as intended' mean, and who decides what use is intended? What can the FDA do if intended use doesn't match actual use?

On the other side of the Atlantic, the EU tests some fragrance ingredients through the European Commission's Scientific Committee on Consumer Safety. But, as in the US, what the scientists conclude can't necessarily be put into action by the regulators, whose powers are limited. The EU does better than most—they have the regulation, mentioned earlier, that restricts twenty-six fragrance chemicals. But, as we saw, the chemical can still be included in a fragrance: it just has to be declared on the label and used in a specified concentration. There are another fifty-six fragrance ingredients that the EU scientists have declared to be 'established allergens in humans' that they haven't yet been able to restrict at all.[8]

Things that cause skin allergies are among the most straightforward chemicals to restrict, because the symptoms can be easily and undeniably linked to the exposure. This is less simple with other health effects. That's why the twenty-six restricted chemicals, and the fifty-six others that the scientists would like to see restricted, are all skin sensitisers, even though fragrance ingredients can produce various other disorders as well.

It's not hard to imagine the pushback from the fragrance industry over every inch of the contested ground of regulation and restriction. The European Commission makes the perhaps wry comment that 'detailed regulation of individual substances used for cosmetics has proven very complex, resource-intensive and difficult to administer.'[9]

—

A cautionary tale about the limits to consumer protection concerns an artificial musk patented in 1959, acetyl ethyl tetramethyl tetralin or AETT, given the trade name Versalide.[10] It was designed to replace some of the earlier musks, which had been found to be carcinogenic. Versalide was a wonderful invention: it had a strong and pleasant musk odour, it didn't discolour when exposed to light, and it was cheap to make. From 1959 onwards it was used in large amounts in all sorts of fragranced products, but especially in laundry detergents.

Then, in 1978, twenty years after it first went on the market, Versalide was tested, apparently for the first time. The rats exposed to Versalide became very ill, very quickly. They developed a blue discolouration and behaved in strange ways—abnormal movements and hyperirritability—that suggested nerve damage. Post-mortem revealed 'structural damage throughout the central and peripheral nervous systems', including 'spectacular myelin bubbling'.[11]

Some time later Versalide was tested again. There were clear signs that even at very low doses the chemical was radically damaging the rats' central nervous systems—their brains and spinal cords. The autopsy found that 'a green substance was formed in the gastro-intestinal tract, and most tissues, including the central nervous system, showed a green-blue or grey colouration.'[12] In 1982, the International Fragrance Association prohibited Versalide from use in any fragrance product.

Further tests showed exactly what kind of damage Versalide did to the brain: it caused degeneration of the myelin sheath—a protective layer that surrounds neurons.[13]

Degeneration of the myelin sheath has severe health effects. It's the characteristic damage of multiple sclerosis. (As it happens, MS has become more common over the last few decades. A review of all the available studies in 2010 found 'an almost universal increase in prevalence and incidence of MS over time...and suggest a general...increase in incidence of MS in females. The latter observation should prompt epidemiological studies to focus on changes in lifestyle in females.'[14])

It's impossible to know how many people were exposed to Versalide over the two decades between 1959 and 1981, and impossible to know how many people suffered health effects from it. Even if the manufacturers of laundry detergent and soap were prepared to say which of their products formerly contained Versalide, who's going to remember what brand of laundry detergent they bought all those years ago?

We're lucky that Versalide finally happened to be tested. Otherwise, it would still be in our laundry detergents, making our sheets smell good but damaging our myelin sheaths. But what about all the other new chemicals out there, including the ones that are now used in place of Versalide? Like the older chemicals, the new ones aren't tested for safety before they're released. Many of them are never tested at all. Yet they're being used in products that we use every day—and we don't have any way of knowing which ones.

~

The extent of government safety regulation these days can feel like overkill. Children's playgrounds now are so boringly safe it's a wonder that any child wants to play in them. My sister-in-law can't sell the eggs from her happy farm chooks without a certificate from the health department. Yet in other ways—ways just as critical to our safety—the protective shield of regulation is strangely absent. We're exposed, every day, to powerful chemicals in fragrance. They're largely untested, mostly unregulated and, in many cases, not declared on the label. Yet, when it comes to those chemicals, we consumers are on our own.

This seemed so strange and so inconsistent that I started to doubt the information I'd found. Surely there are experts who aren't just labelling or assessing fragrance, but are making sure it's safe? Well, yes, there are. The only problem is, they're the same people who make it.

CHAPTER 7

IN DEFENCE
OF FRAGRANCE

In 1973 the perfume manufacturers of the world got together and formed the International Fragrance Association, IFRA. One reason for its formation was that consumers were starting to ask questions about the fragrances they were using every day. According to its website, IFRA 'fosters a sense of understanding of consumers' needs and demands a sense of responsibility in their satisfaction. Adding to our sense of wellbeing. And increasing our sense of prosperity.' Its list of ingredients has been published 'to support our drive for increased transparency'.[1]

IFRA's other reason for existing is that it's in the privileged position of being its own regulator. The only real overseer of the fragrance industry is the fragrance industry itself. As it says on its website: 'The IFRA Standards form the basis for...the self-regulating system of the industry, based on risk assessments carried out by an independent Expert Panel.'[2]

The Expert Panel works with a lab that IFRA funds, the Research Institute for Fragrance Materials. RIFM is described (in the 'Conflict of Interest' section of its published papers) as 'an independent research institute supported by the manufacturers of fragrances and consumer products containing fragrances'. Its main job is to run tests on fragrance ingredients. Some of these studies are published in peer-reviewed scholarly journals.

A lot of IFRA's publicly available information shows that some fragrance components can produce bad health effects. IFRA's own studies, as well as the ones done by independent labs, show that some of those chemicals can cause allergic sensitisation. They also confirm that some are carcinogens.[3] It's IFRA's published studies, among others, which show that certain fragrance ingredients can cause other serious health effects.

So perfume manufacturers acknowledge that some of their ingredients can lead to problems. But they can't stop using all those chemicals. If they did, their palette of fragrance ingredients would dramatically shrink. That would be bad for business: consumers like complex, subtle fragrances containing a lot of different ingredients. They also demand a constant stream of new fragrances.

The industry would take a massive hit if hundreds of ingredients were suddenly banned.

In any case, it probably wouldn't stop dangerous fragrance chemicals being used. Even if all the members of IFRA stopped using the health-compromising ingredients, those chemicals would still be in some products, because not all fragrance manufacturers are members of IFRA and bound by its regulations. And, in a competitive market, some IFRA members might find it worth their while to risk expulsion by breaking the rules.

And consumers might not mind. Perversely, many perfume fans seem willing to overlook the odd spot of carcinogen in their favourite perfume. Some of the most glamorous old perfumes, such as Chanel No 5, originally contained ingredients that have since been banned.[4] Parfumiers have reformulated the fragrance to replace the problem ingredients, but the internet is full of people looking for vintage bottles of the original Chanel No 5 because they don't think the new one is as good as the original. They know why the perfume was reformulated, but they're prepared to take their chances. One fan of old perfumes, aware that the nitro musks used in some of them can be carcinogenic, explains her thinking:

> When people find out I collect vintage perfumes the usual question is 'Do you really wear those old perfumes?' Well of course!...I too have had some minor allergic reactions but nothing to deter me. I've

also read the warning on some of the ingredients such as the nitro musks...I say, experiencing perfumes that haven't existed in decades, from bygone days, is worth the risk.[5]

—

So the fragrance industry has to find a way to reconcile two opposing facts: the health risks of some of its ingredients, and the impossibility of not using them.

IFRA has spent a great deal of money, time and effort to find a way through this dilemma, and it's come up with a compromise called a Quantitative Risk Assessment. This is a standard type of scientific investigation where researchers set out to find exactly what *quantity* of a substance puts you at *risk*—how much you can have before you get sick. This kind of assessment starts from the well-accepted assumption that there can be a safe dose even for a dangerous chemical. Chemists often quote the sixteenth-century doctor Paracelsus: 'The dose alone makes the poison'—in other words, there's a safe dose for any substance, no matter how toxic. The job of a QRA is to put a number on that safe dose.

Different doses of the chemical are given to different groups of lab animals. The animals given the largest doses may show adverse health effects, up to and including death. The animals given the lowest dose will be expected not to show any effects. Somewhere in between is the largest dose the animal can be given that doesn't make it sick. This is the No Observed Adverse Effects Level, or NOAEL. That's the maximum safe dose for that particular animal.[6] (RIFM uses a variety of lab animals, because different animals respond differently to the chemicals being tested. Rats, mice, hamsters and rabbits are commonly used.)

The studies on fragrance ingredients at RIFM use Quantitative Risk Assessment to test for a variety of health problems. Some test whether the chemical will produce an allergic reaction on the skin. Some are toxicology tests, which find out how poisonous the chemical is. A few chemicals are put through a two-year carcinogenicity study.

Taking these results as a starting point, the RIFM chemists calculate a NOAEL for a human. The IFRA website doesn't go into detail about exactly how this extrapolation from animal to human is arrived at—that information is only available to IFRA members. This is a pity, as the whole area of animal-to-human extrapolation is something scientists disagree about, and it would be helpful to know how IFRA arrives at its conclusions.

But, in the spirit of transparency that it promises, IFRA does publish on its website two lists of fragrance ingredients that its tests have shown to be significant health risks for humans, and which it prohibits or restricts.[7]

The 'prohibited' list names fragrance chemicals that

were used in the past, but are now considered too danger-ous. IFRA regulations state that they can't be used in any amount, in any product. There are seventy-six ingredients on this list.

The 'restricted' list names fragrance ingredients that have been shown to cause health problems, but are still allowed to be used, as long as the concentration is lower than a certain percentage. There are 105 chemicals on this list.

Putting both the lists together gives a total of 181 chemicals that are subject to some form of regulation. Of the total number of fragrance ingredients—2,947, accord-ing to IFRA—this is a little over six per cent.

The health effect—what IFRA calls the 'critical effect'—for each restricted substance is specified. The critical effect for most of the chemicals on the list is sensitisation: they can cause skin allergies. Others are phototoxic—light can cause them to damage skin. Four restricted ingredients are named as having the 'potential for nitrosamine formation' (most nitrosamines are carcinogenic).[8] Four are listed as having the critical effect of being carcinogenic: benzene, furfural, methyl eugenol and estragole. Of those four carcinogens, two—benzene and furfural—are listed as 'not to be used as fragrance ingredients'.

This form of words is carefully chosen. Benzene and furfural aren't generally added to fragrance as an intended ingredient—but they may be in there, as a contami-nant. The other two carcinogens—methyl eugenol and

estragole—can still be used in fragrance, but only in specified concentrations.[9]

—

The next challenge for IFRA's lab is to find a concentration of each restricted chemical that will be safe for
consumers. They could simply say: this chemical mustn't
ever be used above such-and-such a concentration. But
some products result in more exposure than others, and
fragrance makers want to use the maximum safe amount.
So IFRA has worked out a fine-grained way of calculating
a safe dose for different kinds of products.

Their rule of thumb is that the more a product is
in contact with the skin, the more the consumer will
be exposed to the fragrance in it. So a product that's
rinsed off (like a shampoo) will result in less exposure
than a product that's left on the skin (like a moisturiser).
Something that you use on your whole body (like a body
lotion) will give you a bigger dose than something you
use only on your face.

IFRA has come up with eleven categories of products
and given each one a Sensitisation Assessment Factor.
To arrive at that SAF, the scientists do a calculation
that combines two elements: how much the consumer is
exposed to the product, and how health-compromising the
chemicals in it are. Again, exact details about these important calculations are only available to IFRA members, so

it's not possible for the public to see how they're worked out, or to debate the figures. But the general picture is that the higher the Sensitisation Assessment Factor of a product is, the more health-compromising it could be.

The highest SAF is 300. Products in this category include Category 1 (things like lipsticks, where the product might be ingested), Category 4 (perfumes and colognes, where the fragrance chemicals are a high proportion of the product) and Category 7 (intimate wipes and baby wipes, where the product is in contact with especially sensitive skin). Products with an SAF of 100, in the mid-range of risk, include Category 5 (hand and face cream, used on a limited skin area) and Category 9 (rinse-off products like shampoo).

Category 11 has the lowest SAF—only 10. This category includes all kinds of room fragrances: toilet blocks, air fresheners, incense, reed diffusers, potpourri, scented candles, and the various air-circulating devices that produce 'ambient scents' in hotels and shops. These products are generally allowed to contain any amount of any chemicals, even ones that are restricted in other kinds of products. This is because there's usually no direct skin contact with them. As the IFRA website puts it, 'due to the expected negligible skin exposure from such products the risk of induction of dermal sensitization through the normal formulation and use of such products is considered to be negligible. As such, the concentration of fragrance ingredients is not restricted in the finished product.'[10]

The exception to this is methyl eugenol. Its carcino-genic potential is so high that IFRA restricts its use (to 0.01 per cent) even in a room fragrance such as an air freshener.[11]

There's one kind of product where IFRA acknowledges that no amount of exposure is safe: things made to go in children's mouths. 'IFRA prohibits the use of fragrance materials and mixtures in toys or other children's products where there is the likelihood of mouth contact.'[12]

—

IFRA doesn't deny any of the bad news about fragrance ingredients. But it's got to find a way its members can go on using them. The RIFM scientists are confident that the concentrations arrived at by their Quantitative Risk Assessment represent a safe dose for consumers. This is why IFRA can say: 'although some substances in common use today may have the potential to cause dermal sensitization, they can still be formulated into consumer products at safe levels.'[13] Their confidence in setting a safe level even extends to substances that they acknowledge have the potential to cause cancer.

IFRA's faith in RIFM's calculations is admirable. It's just a shame that this reassurance isn't as independent as it might be: after all, RIFM is supported by the fragrance industry. There's another problem, too. IFRA may be right that, used in the concentrations laid out in its regulations, its products are safe. But that's in the artificial world of those tests, where lab animals are exposed to one chemical at a time, in precise doses, on square millimetres of skin. But what about the messy real world, where the rest of us live?

CHAPTER 8

BEYOND THE LAB

It's early morning and you're in the bathroom getting ready for the day. You reach for the shampoo and squirt out a blob. The label says to keep lid closed but you're in a hurry and leave the little flicky thing open. Then the conditioner. You're supposed to rinse it out but you've found it works better on your curls if you leave it in. Face cream: well, it's called that, but you use it on your hands as well. Deodorant: if a bit is good, won't more be better? Finally, the favourite scent: do the pulse points, but let's do a few good squirts all over for good measure.

The RIFM scientists have made their scrupulous calculations about 'intended or reasonably foreseeable use'. But they're not in the bathroom with you. There's no one to tell you that one reason you're supposed to close the shampoo bottle is that otherwise some of the chemicals in it may react with air and produce a carcinogen. No one is there to explain that the safe dose of conditioner has been calculated on the assumption that it's going to be rinsed out. Someone in a white coat has smeared the face cream on a precisely measured area of lab-rat skin—not a millimetre more or less—calculated the average area of the human face and, on the basis of that, come up with a safe dose of the chemicals in the cream. But you've put it not just on your face but also on your hands and anywhere else you thought needed moisturising.

There are a lot of differences between you and that lab rat. One is that it was sitting in its cage with the face cream on a patch of skin, but that was the only fragrance it was exposed to. It didn't have other scented products on that bit of skin as well. How's a researcher expected to know all the different products you've chosen to use that morning? All the white coat can do is test one chemical at a time, record the results and do the mysterious maths to extrapolate to a human. But if you've used, say, up to ten fragranced products in your bathroom half-hour, and each of those products contains, say, a couple of dozen fragrance chemicals, that's a mix that the white coat can't possibly anticipate. The human body can perform marvels in detoxifying itself, but it has to work a lot harder, faced with dozens of chemicals, than the lab rats did when they were given one at a time.

Actually, chances are that a lot of those products will contain some of the same chemicals. The soap, the shampoo, the conditioner, the face cream and the perfume are all quite likely to contain the restricted ingredient citronellol (a nicely lemony-smelling chemical) as part of their mix. The researchers have stipulated what they consider to be the safe dose of citronellol for all these products. But if you're simultaneously getting the safe dose in six different products, you're getting six times the safe dose. (And that doesn't count all the other sources of citronellol that you'll be breathing that day from everyone else's shampoo and perfume.) RIFM tries to account for this by adding a margin of safety on top of their calculation of the 'safe exposure'. But there's no way of knowing whether that reflects what happens in the real world.

So there's an accumulation effect. There's also an interaction effect. Some innocent chemicals, when you put them together, react to make something dangerous. A familiar example is bleach and ammonia. The label on each of these warns you not to use them together, because they combine to make toxic chloramine gas. Fragrance makers know that chemicals can interact, and *within* each fragranced product they'll make sure it won't happen. But they can't control what happens *between* products. How were they to know that you were going to use *this* face cream, and then squirt *that* perfume on?

Then there's the reacting-with-air problem. Some

fragrance chemicals are safe when the manufacturers put them in the bottle, but react with air as soon as you open the lid—as we've seen, the common fragrance ingredients limonene and pinene are an example. In contact with air, they produce the carcinogen formaldehyde. This is why labels tell us to 'keep bottle tightly closed'. But, even if we obediently do that, once the limonene is out on the skin it's exposed to the air and will start oxidising.

The IFRA regulations are pernickety when it comes to anything we put on our skins. But skin contact is only one way that fragrance chemicals get into our bodies. The whole point of a fragranced product is that we can smell it, and the reason we can smell it is that the fragrance molecules are floating freely through the air. From there they go up our noses and into our brains. They also land on our skin, and some of them are able to pass through it into our bloodstream. They travel into our lungs, too, and get into our bloodstream from there.

Because we don't rub room fragrances directly on our skins, IFRA has decided that these products are generally exempt from its restrictions. Most Category 11 products— all those room-scenting devices—can contain unlimited amounts of fragrance chemicals, even those restricted in every other kind of application. Yet the chemicals from air fresheners and potpourri and incense enter our bodies just as surely as the chemicals from shampoo or face cream do. It's an enormous blind spot in IFRA's safeguards. As the researcher Anne Steinemann comments: 'If it's coming out of a smokestack or a tailpipe it would be regulated, but if it's coming out of an air freshener [it's] not regulated.'[1]

—

But can fragrance chemicals really get through skin? It's been found that they can, some more easily than others.[2] Beyond the skin they can travel into the bloodstream, to be carried all around the body.

Until recently, the fragrance industry's calculation of skin absorption erred on the side of safety. Unless there was specific information about how easily a chemical could penetrate the skin, IFRA regulations assumed the degree of absorption to be one hundred per cent—that is, they assumed the worst. That gave a wide margin of safety.

But, in 2014, researchers set out to measure the *actual* skin absorption of various fragrance chemicals rather than assuming a one-hundred-per-cent default, which they described as 'unreasonable'. (This investigation 'was part of an internal research program of the Research Institute for Fragrance Materials, Inc.'—in other words, it was funded by fragrance manufacturers.) If skin absorption were less than one hundred per cent, that would affect the safety assessment of a fragrance ingredient and, as this study puts it, 'may bring the total systemic exposure below the Threshold of Toxicological Concern leading to different risk management'. In other words, if the sum were done differently, it's likely that fewer products would need to be on the restricted list.[3]

Their research sorted fragrance ingredients into three levels, depending on the degree of skin absorption. The lowest is ten per cent, the middle is forty per cent and the highest is eighty per cent. This allows IFRA to downgrade the safety rating of many chemicals, since even the highest absorption rating is less than one hundred per cent. In some cases the safety rating can be downgraded radically, from a hundred per cent to ten per cent. Whether this 'different risk management'—reclassifying chemicals as safer—will actually make things safer for consumers is another question.

—

What of that worthy regulation about not using fragrance in products for children if there's 'likelihood of mouth contact'? The fragrance manufacturers are right that this is a good idea: children's bodies are much more vulnerable than adults'. Their immune systems—where the body deals with threats to its health—are still developing and can be easily confused or overwhelmed. The organs that deal with toxins, like the liver and kidneys, aren't yet running at full efficiency. A newborn's blood–brain barrier is more permeable than that of an adult. Infants have more surface area of skin relative to their weight than adults do. And their skin is 'poorly keratinized'—the protective outer layer hasn't fully formed yet, so a newborn's skin can absorb around three times more than an adult's.[4] All this adds up to a double whammy—more skin to absorb chemicals, but less-developed ways to get rid of them.

When did those regulators last meet an actual child? As far as little children are concerned, there is *no* object where there isn't 'likelihood of mouth contact'. Children put everything in their mouths. Mum's finger, with its little load of fragranced hand cream. Blankie and Teddy, clean-smelling from fragranced laundry powder. Rubber Duckie foaming with fragranced bubble-bath water.

And who needs mouth contact? Skin contact will do it, too, especially with that 'poorly keratinized' baby skin. The nappies, the baby wipes, the nappy cream, the talcum powder, the shampoo, the bubble bath—traces of the fragrance chemicals in all of those will remain on the child's skin, and from there they can be absorbed into the bloodstream. They'll be in the air, too, being breathed into the child's immature lungs.

The thing that makes a mockery of this concern for children is that IFRA members now sell perfume specifically for babies. (When someone told me this I scoffed, but when I went online I found that, incredible though it seems, it's true. Many of the major brands now sell baby fragrances.) Yes, it may be the case that there's a low 'likelihood of mouth contact' there. But in health terms that's irrelevant. The chemicals in those baby perfumes will easily go through the baby's soft, absorbent skin into his or her bloodstream. From there those chemicals will travel to every organ in the body: the liver, the heart, the lungs, the eyes, the brain. These are still developing. How can people think of saturating them with chemicals that include irritants of all kinds, sensitisers, toxins, carcinogens and hormone disruptors?

—

The tests that RIFM does are properly run scientific studies. Their methods are well accepted, their results reliable, and they're published in peer-reviewed scholarly journals. Still, like many scientific studies, they have some significant limitations—and consumer-protection regulations built on scientific studies can only be protective if the science of those studies is watertight.

A major weakness is that not all of the nearly three thousand fragrance chemicals have been tested. Testing is time-consuming and expensive. Even a basic toxicological test is a large-scale affair involving many lab animals over several weeks or months. Testing to see if a chemical is a carcinogen is even more expensive. A two-year cancer study using lab animals costs around US$2 million per test.[5]

The most straightforward kind of test is for skin allergies—and this is one reason why IFRA's regulations are mostly about those complaints. But skin contact is only one way of being exposed to fragrance chemicals, and skin allergies are only one kind of health problem.

Testing is expensive, but it's not compulsory. In fact, the question really isn't: why aren't all chemicals tested? It's more like: why are any chemicals tested? Only IFRA knows how many chemicals its lab has tested. And, although many of its studies have been published and are freely available, there may be others where the results haven't been made public. No lab is obliged to make its results public, even if it would be in the public interest to do so.

These realities are reflected in the limits of the data available from RIFM studies. Many fragrance chemicals have been tested for skin sensitisation. Fewer have been tested for toxicity. Fewer again have been tested for carcinogenicity. And for many chemicals, as we've seen, there's no data at all.

If there've been no tests (or none that have been published), there'll be no negative reports, no matter how harmful the chemical may be. Scientists, fond of a pungent aphorism, like to remind us that 'absence of evidence does not mean evidence of absence.'

Even the longest studies on fragrance chemicals may not go on long enough. Cancer can take decades to develop. Hormone problems can take even longer—they can begin before birth, via the mother's bloodstream. The long-term animal studies don't necessarily reflect the real lifetime exposure that humans have between the womb and the nursing home.

A significant limitation of IFRA's tests—or at least the regulations based on them—is that the calculations about safe exposure for humans are mostly based on animal studies (the exceptions are a few skin-allergy studies on human subjects). Rats and mice are mammals like us. They get tumours as we do. Their brains, like ours, are affected by chemicals. But many of their body mechanisms are quite different from ours. Animals have evolved to be more sensitive to some chemicals than we are, but less sensitive to others. Some of the ways they detoxify chemicals are different from ours, too.

This means that something that causes problems in an animal may not cause them in humans. We can eat chocolate with no ill effects other than a pang of guilt, but chocolate can kill a cat or dog. It works the other way, too: something that causes no harm to an animal may cause serious damage to humans.

—

The IFRA Code of Practice says, 'IFRA members must comply with IFRA standards.' To police that 'must' the organisation runs a compliance program, which tests a sample of fragranced products every year to check that the manufacturers aren't using more of a dangerous chemical than they should be, or using a prohibited one.[6]

Has IFRA ever found a non-complying manufacturer? Its website doesn't say. But other, independent tests have found many cases of fragrance products containing prohibited chemicals, a greater-than-allowable amount of restricted chemicals and ingredients not declared on the label when by law they should be.[7]

If IFRA did find a member who was ignoring the regulations, how would it make that member toe the line? The only power it has is to expel the company, depriving it of the advantages of membership. But there's big money in fragrances, and some of those restricted chemicals smell very good. A fragrance manufacturer might do the sums and decide it was worth being expelled, if it meant it could get a jump on its competitors.

IFRA knows that consumers are starting to be concerned about fragrance. It's working hard to convince us that its regulations will keep us safe. But IFRA, while announcing its commitment to consumer safety, covers itself with a disclaimer: 'The responsibility for the safety of any fragrance material put on the market rests with the company supplying the material.'[8]

So behind all the fragrance manufacturers' worthy rhetoric, and their elaborate and expensive superstructure of self-regulation, is an unfortunate fact: essentially, it's all wishful thinking. Even if IFRA's regulations were completely effective in ensuring consumer safety, IFRA has no real power to enforce them, no final responsibility for them and no accountability to anyone else.

But does that matter? Is fragrance something we need

to be protected from? Asthma can be life-threatening, true. Headaches and rashes spoil a person's day. But in a world full of dangers and threats to our health, are a few headaches or asthma attacks really worth worrying about?

CHAPTER 9

AFTERSHAVE WITH WHAT?

Dr Luca Turin is a biophysicist who specialises in the sense of smell and the chemistry of fragrance. He's the author of several standard reference books about perfumes. He can be watched doing an entertaining TED talk.[1] He mentions coumarin, a common fragrance chemical that has a pleasant 'new-mown hay' sort of smell. Coumarin occurs naturally in a lot of plants, and in either its natural or synthetic version it's been widely used in perfumes and cosmetics for at least a century. Turin says coumarin has been used in men's colognes since 1881. 'The only trouble with it'—he pauses—'it's a carcinogen, so nobody likes...you know... aftershave with carcinogens...' The audience erupts into laughter, thinking this wryly amusing man is making another joke.

Actually, you can see he's a bit surprised by the response. Like all fragrance scientists, Turin knows that some fragrance ingredients are carcinogenic. Coming up with less-toxic alternatives is one of the things that keeps researchers like him in business. As he puts it in the talk, 'They asked us to make a new coumarin.' But for the audience, the idea that a fragrance ingredient could be dangerous was absurd. Aftershave with carcinogens! Very amusing.

Luca Turin is right about coumarin being a carcinogen—for some time it's been known to cause cancer in lab animals.[2] Since then, scientists have been trying to establish whether or not it causes cancer in humans too. So far they're still not certain, because there haven't been enough studies to settle the question. But, as the European Commission's Scientific Committee on Food put it, 'the new data on liver metabolism did not reassure the Committee.'[3] The International Fragrance Association has put coumarin on its list of restricted fragrance chemicals.

Coumarin is absorbed readily through the skin.[4] If you're exposed to a fragrance containing coumarin, you may be absorbing (through skin and lungs) something that is definitely a carcinogen for lab animals, and hasn't been ruled out as a human carcinogen, too.

Not all those lab animals got cancer when they were exposed to coumarin. In the same way, some smokers beat the odds and live to be a hundred. But the way carcinogen studies work is by looking at large numbers of animals, and comparing the ones that are exposed to the chemical with the ones that aren't. Scientists are reluctant ever to say that one thing 'causes' another, because that's difficult to prove beyond doubt. But if more of the exposed group than the unexposed group gets cancer, the chemical will 'correlate' with cancer—there'll be a pattern showing an association between the chemical and cancer.

Scientists are always careful to point out that 'correlation is not causation.' Two things can go together without one causing the other. But, in layman's terms, if a chemical has been classed as 'carcinogenic' it means that, if you're exposed to it, your chances of developing cancer are greater than if you're not exposed.

—

It turns out that coumarin is one of a number of fragrance ingredients that are carcinogenic in animals. Some of them are among the ingredients most commonly used.

When Anne Steinemann analysed fragranced products for her study of 'sick building syndrome', one of the most commonly occurring chemicals was benzyl acetate. It was in half the products in her first sample, and in over a third in the second sample. Benzyl acetate has been shown to be carcinogenic for animals.[5]

Beta-myrcene is another very common fragrance ingredient.[6] Researchers have found it associated with kidney cancer in rats.[7] The National Toxicology Program in the US concluded that beta-myrcene showed 'clear evidence of carcinogenic activity in male F/344N rats'.[8]

Isoeugenol is a chemical with an attractive sweet, spicy odour. It's a common ingredient in fragrance: one analysis of bottled perfumes revealed that isoeugenol was present in over half of them.[9] When the National Toxicology Program tested it on lab animals, they found 'clear evidence of carcinogenic activity' in the liver, blood, nose, mammary gland and thymus (part of the immune system), as well as damage to kidneys and stomach.[10]

Allyl isovalerate has a pleasant cherry or apple smell and is used in many fragranced products. The National Toxicology Program tested it on experimental animals and concluded that it was carcinogenic, producing leukaemia and lymphoma.[11]

Estragole is a chemical with an aniseed smell that makes it a popular fragrance ingredient. Researchers for the National Toxicology Program tested estragole on lab animals in 1999 and concluded that it was a carcinogen.[12] A few years later the European Commission's Scientific Committee on Consumer Safety did a review of all the previous data and concluded that it was not only carcinogenic but also genotoxic: it damages the genes, causing mutations. 'Estragole has been demonstrated to be genotoxic and carcinogenic. Therefore the existence of a threshold cannot be assumed and the Committee could not establish a safe exposure limit. Consequently, reductions in exposure and restriction in use level are indicated."[13]

~

Well, those poor rats and mice. But what about humans? Is any of this stuff going to harm us?

IFRA's researchers are prepared to say no, not in the concentrations they recommend. All the chemicals named above are permitted for use in fragranced products. Benzyl acetate, beta-myrcene and allyl isovalerate can be used in any product, in any concentration. Isoeugenol and estragole can only be used in specified concentrations in skin-contact products, but can be used in any amount in air fresheners and other room fragrances.[14]

IFRA may be prepared to name a safe level for these chemicals, but other scientists—researchers not funded by the fragrance industry—aren't. Independent researchers are clear that the science doesn't justify assuming that these chemicals are safe for humans.

The International Agency for Research on Cancer is one of the World Health Organization's research bodies. It assesses chemicals where there's evidence of carcinogenicity in animals, to see whether the chemicals are likely to be carcinogenic to humans, too. One of its categories is for chemicals that show carcinogenic activity in animals, but where there's little or no data for their effect on humans. Each of the chemicals named above is in that category:

'not classifiable as to its carcinogenicity to humans'.[15] The World Health Organization isn't prepared to declare any of them safe for humans, no matter what the concentration. All it's prepared to say is that no one knows.

The reason is a lack of information: there haven't been enough studies. And there haven't been enough studies because doing definitive carcinogen studies on humans is more or less impossible. It's surprisingly hard to find human volunteers who are willing to sit in a room being exposed to coumarin or estragole for a couple of years while researchers monitor them for signs of cancer, and who are then prepared to be euthanised and have their bodies dissected.

Working the other way round—studying people who've already got cancer, and trying to work out how it started—is tricky, too. Cancer generally takes years to develop, and during those years a person is exposed to thousands of different substances that might have contributed to the disease. It's not often possible to find a smoking gun—one unmistakable cause.

But there are a few ways in which scientists can investigate whether a fragrance chemical is carcinogenic for humans as well as animals. One is to study exactly how the cancer develops in the animals. Sometimes this pathway is a bit of biological machinery that humans don't have. But sometimes it's the same. If cancer develops via the same process in humans and in that particular animal, then there's a high likelihood that the chemical is carcinogenic to both.

Researchers can also get information through epidemiological studies. These are number-crunching analyses of large numbers of people, looking for patterns. If a group of people all get sick in the same way, and have something in common—for example, if they all work in the same industry—that common element might have something to do with them getting sick. It was an epidemiological study that established the link between smoking and cancer: over forty thousand British doctors were studied for nearly fifty years. Those who smoked had a higher chance of getting lung cancer than those who didn't.

Epidemiological studies on fragrance chemicals are tricky to do for a couple of reasons. One is that we're all exposed to fragrance chemicals—there's no way to run a control group of people who aren't exposed. The other is that it's impossible to pull fragrance exposure out of all the other chemical exposures in our lives. A person can probably tell you how many cigarettes they smoked in a week, but couldn't tell you how many millilitres of estragole they'd been exposed to.

—

But, in spite of the shortage of human volunteers, and of epidemiological studies, there is some data about whether certain fragrance ingredients are carcinogenic for humans.

The National Toxicology Program maintains a Report on Carcinogens—the top level of this register is chemicals that are 'known to be human carcinogens'.[16] Given the difficulties of saying for sure that something is carcinogenic for humans, being in this top level is quite a distinction, enjoyed by only about fifty substances, including tobacco smoke, mustard gas and asbestos. Formaldehyde is one of them.[17]

Formaldehyde is sometimes added to fragrance as an intended ingredient—it's a useful preservative. But one of the reasons it's emitted by so many fragranced products is that it's formed from some of the chemicals that go into fragrance. As we saw earlier, limonene and pinene were the most common fragrance ingredients in Professor Steinemann's 'sick building' studies. They're both terpenes, and terpenes produce formaldehyde when in contact with air.[18] So does the common fragrance preservative quaternium-15.[19]

Along with formaldehyde, acetaldehyde was among the 'most-common ingredients' found in both of Steinemann's studies. IARC has listed acetaldehyde as a 'probable human carcinogen'.[20] The National Toxicology Program includes acetaldehyde on its list of substances that are 'reasonably anticipated to be a human carcinogen'.[21]

The European Commission's Scientific Committee on Consumer Safety assessed acetaldehyde in 2012, and confirmed that it was carcinogenic to animals after both oral and inhalation exposure. The committee also noted that acetaldehyde is a skin, eye and respiratory-tract irritant, and that it induces gene mutations in experimental animals. It concluded that acetaldehyde 'should not be used as an intended ingredient in cosmetic products except...at a maximum concentration of 0.0025% (25 ppm) in the fragrance compound'.[22] Acetaldehyde is on IFRA's list of fragrance ingredients, but it isn't on the restricted list, so it can be present in any concentration in any fragranced product.

Another chemical that isn't deliberately added to fragrances, but is often emitted by them, is 1,4-dioxane.[23] That's because it's a common contaminant. On the National Toxicology Program's Register of Carcinogens it sits with acetaldehyde on the second level: it's 'reasonably anticipated to be a human carcinogen'.[24]

Apart from this danger, 1,4-dioxane is a skin and eye irritant, and causes irritation to the respiratory tract. Exposure may cause damage to the central nervous system, the liver and the kidneys. It can be absorbed by inhalation and through the skin.[25]

The Food and Drug Administration in the US encourages manufacturers to remove 1,4-dioxane from their products, but they aren't required to. It's not on IFRA's list of prohibited or restricted substances, so 1,4-dioxane can legally occur in any concentration in any fragranced product.

Methylene chloride (otherwise known as dichloro-methane) is another fragrance ingredient classified by the National Toxicology Program as a 'substance reasonably anticipated to be a human carcinogen'.[26] It's a carcinogen in experimental animals, and was found to produce an excess risk of multiple myeloma (a bone-marrow cancer) in humans.[27] Methylene chloride isn't prohibited from use in fragrances, and it isn't restricted, so it can be used in any amount in any product.

Methyl eugenol has a pleasant smell described as 'a clove-carnation odour' and it's widely used as a fragrance ingredient. The National Toxicology Program describes it as a 'substance reasonably expected to be a human carcino-gen'.[28] IFRA acknowledges on its website that methyl eugenol is carcinogenic.[29] In fact, as we saw, it considers the chemical to be so unsafe that it even restricts its concentration in 'non-skin products' such as air fresheners, which are gener-ally unrestricted.

This list of carcinogenic fragrance ingredients isn't necessarily complete. More digging in the scientific literature could well turn up more. And as well as these known fragrance carcinogens, there's an unknown number of others: all the fragrance materials that haven't been tested, or where the test results haven't been published. When you research fragrance chemicals, you often find yourself reading entries such as 'Carcinogenicity: no data was available'.

—

None of this information about cancer-causing fragrance ingredients would surprise the fragrance manufacturers. But, as we saw, they maintain that their Quantitative Risk Assessments ensure that the amount of carcinogens in fragrance is small enough to be safe. In other words, it's not a question of whether the ingredients are carcinogenic, but whether they're carcinogenic *enough*.

So how much of a carcinogen does it take to give you cancer? The only sure answer is that it's a different amount for everyone. It depends on your genes, your lifestyle and the environment you live in. The IFRA-funded researchers are confident of the limits they put on those carcinogens, but how can they be so sure?

For a start, no one knows what the safe dose of any carcinogen is for a human. To know that, you'd have to do those impossible carcinogenicity studies on people.

But, even if you did know what dose was safe, it wouldn't help consumers. As we've seen, the ingredients of fragrance in a product aren't disclosed, and the amount we get from second-hand sources is impossible to calculate. So it's true that there may be a dose for formaldehyde, or 1,4-dioxane, that won't trigger cancer. But it's impossible to know what that dose is, or that we're on the right side of it.

Some of the makers of fragranced products have evidently come to the same conclusion. A press release by Johnson &

Johnson in 2012 announced that the company was planning to reformulate the products it made for babies so that those products didn't emit the carcinogens formaldehyde and 1,4-dioxane. It also promised to remove formaldehyde from products it made for adults, except in 'exceptional cases'. By 2014 it declared that its baby products were free of formaldehyde and that the levels of 1,4-dioxane in them were 'the lowest measurable dose ever'.[30]

This is good news, but it's also the mirror image of the bad news. Who knew, until that announcement, that Johnson's Baby Shampoo—an emblem of all things safe—had contained at least two carcinogens presumably for over sixty years, ever since it was formulated in 1953?

Other manufacturers and retailers (including Procter & Gamble, Walmart and Target) are also making moves towards ensuring personal products are free of formaldehyde and other health risks.[31]

All these manufacturers are responding to the science, but they're also responding to consumer awareness. There's a steady stream of articles in the media about the possible dangers of fragrance ingredients[32] and the 'toxic chemicals in breastmilk!' story is now something of a staple.

It's great that these companies are trying to make their products safer. But what about all the other manufacturers and retailers who aren't so responsible? How will consumers of their products know if there's formaldehyde in the baby shampoo or bubble bath or nappies they're buying?

~

Defenders of fragrance like to point out, correctly, that many of these carcinogens are also found in herbs and spices. There's estragole in basil, for instance. Does that mean we should stop eating it?

There's no way of putting a number on the amount of estragole in our daily exposure to fragrance, so it's impossible to say how much basil we'd have to eat to get the same amount. But there's a natural cut-off mechanism with humans and food. If you eat a lot of a certain food (for instance, enough basil to give you a big dose of estragole), you get sick of it and stop eating it. Humans and basil have been living together for a long time—long enough to evolve that *stop* signal. This isn't the case with fragrances, because—even if they contain some of the same chemicals that herbs do—the whole package is chemically very different, and not one our bodies have evolved to handle.

~

The carcinogen factor puts fragrance in a different category from products that might give a few unlucky people a headache or asthma or a rash. Those symptoms come on quickly, so it's easy to see a pattern and avoid the fragrance. That's what your body is telling you to do.

You could say that those of us who get symptoms straight away from fragrance are the lucky ones. As much as we can, we avoid what's making us feel bad. But people who don't get immediate symptoms go on using fragrance for years. If there's a substance in there that's doing them harm, they won't know until, perhaps, it's too late.

Many different factors have to come together for cancer to start. Who can say which one tipped the balance? And fragrances are only one of many sources of carcinogens in our lives. Still, we go to a lot of trouble not to be exposed even to trace amounts of other carcinogens, like asbestos and radiation and tobacco smoke. If a product is carcinogenic, it seems nothing more than common sense to avoid it if we have a choice in the matter.

CHAPTER 10

THE
INDESTRUCTIBLES

The traditional fragrance chemicals have been around for a long time. But some—artificial musks, phthalates and ultraviolet-light blockers—are recent inventions. Only one human lifetime has passed since they started to be widely used.

When they first appeared, some of these new chemicals were put through the usual tests. Did they cause skin allergies? Were they toxic? Did they cause cancer? Some of the earlier musks—for example, the nitro musks referred to by the vintage-perfume fan mentioned earlier—turned out to be carcinogens, and were replaced, but some of the new chemicals passed the old tests reasonably well.

It took a while, but gradually researchers realised that they might be testing for the wrong things. As so often in

science, the story started with people who went looking
for one thing, but found something else.

—

The point of a wastewater-treatment plant is to break
down the substances that go into it—sewage, and the
grey water from laundries and kitchens and bathrooms.
The aim of the treatment is to make the water as clean as
possible, so it can be put back into circulation.

It turned out that the treatment process was good
at breaking down bacteria and old-fashioned soaps. But,
starting around the 1990s, the scientists who check on
wastewater began to notice an odd thing: some of the new
laundry-powder fragrances weren't being broken down at
all. Grey water was pouring out of the laundries of the
world and the musks in it were whipping straight through
the wastewater-treatment plants. They could still be found,
unaltered, downstream from the plants, in 'apparently
ubiquitous distribution' and in 'not negligible amounts'.[1]

Ever since they realised this, scientists have been
working on ways to improve wastewater treatment to deal
with musks. The scholarly literature is full of their efforts,
but it's clear that they haven't yet fully succeeded. Some
treatment plants remove the musks better than others,
depending on the exact process used.[2]

Musks could also be found in all the things that lived
in the water that came out of the treatment plants. They

were in the flesh of the fish downstream of the treatment
plants, and near the ocean outfalls.[3] They were in drinking water that was drawn from rivers.[4] Even if the water
supply wasn't downstream from a treatment plant, musks
could still be found in it, because the powdery residue of
the chemicals was picked up by the wind, and fell later in
snow and rain.[5] In some studies, virtually every sample
of rainwater that was tested contained synthetic musks.[6]
The musks even got into the carrots and animal pasture
that had been fertilised with sludge from the wastewater
plants.[7] Researchers started to realise that these chemicals were pretty much indestructible.

A substantial body of research confirmed that fish
living downstream of treatment plants were storing the
musks in their bodies, a phenomenon called bioaccumula-
tion.[8] When a tiny fish was eaten by a bigger fish, the
musks in the tiny fish stayed in the bigger fish. An even
bigger fish that ate the big fish got all the musks from
the big fish, plus all the musks from the tiny fish. The
further up the food chain you went, the more the musks
bioaccumulated.

When researchers decided to measure these persistent
musks in humans they found them there, too. Nearly
everyone had them in their bodies—it was common to find
musks in over ninety per cent of the people tested.[9] One
study compared the amount of artificial musks in younger
and older women. The older the women were, and the

more fragranced products they routinely used, the more musks they had in their blood.[10]

They found that toddlers get about five times as much exposure as adults.[11] One reason may be that they spend a lot of time at floor level, among the dust, sticking things in their mouths, because it turned out that house dust contains significant levels of musks.[12] Children and babies were full of the fragrance blender DEP, too: eighty per cent of the infants in one study had DEP by-products in their bodies. The younger the child, the more he or she had. And the more lotions, powders and shampoos he or she had been exposed to, the higher the amount was ending up in the children.[13]

—

Once they get into an animal, musks make a beeline for fatty tissue, and store themselves there.[14] Breasts are mostly fat, and so is breastmilk. Humans are at the top of the food chain, but right at the very pinnacle is a breast-fed baby.

Many studies confirmed that breastmilk commonly contains artificial musks. Moreover, the more fragranced products a woman uses, the more musks there are likely to be in her breastmilk. In a Swedish study from 2008, samples of breastmilk were tested and the women completed questionnaires about their use of fragranced products. The study found that 'Women with a high use of perfume during pregnancy had elevated milk concentrations of HHCB [a synthetic musk], and elevated concentrations of AHTN [another synthetic musk] were observed among women reporting use of perfumed laundry detergent.'[15] As a result, babies whose mothers used more fragranced products were getting a bigger dose of musks than the babies of women who used fewer of them.

There was something unsettling about the thought of babies drinking quantities of artificial chemicals from a source that's a byword for everything sweet and safe. But few studies had been done on any possible adverse health effects that musks might cause. Then, in the early 2000s, studies began to appear that showed a curious thing about artificial musks, something that the chemists who'd invented them hadn't intended or foreseen: although these chemicals don't exist in nature, the body responds to them as if they're its own hormones.

—

Hormones pretty much run the show, biologically. Glands all over the body make various hormones that signal (along with genes) all kinds of things: whether to be fat or thin, whether to feel tired or energetic, whether to grow tall or short. Hormones affect the way the brain develops in unborn children. Hormones are bound up with the immune system. And they're involved with the diabetes and obesity problem. (Farmers know about the growth-enhancing effects of hormones on their cows and chickens, and scientists have coined the word 'obesegens' to describe certain chemicals that alter the way fat is metabolised in the body, or alter appetite.)

The signal doesn't have to be enormous: just a little bit is often enough. A familiar example is the contraceptive pill. The amount of hormones in the pill is tiny, yet it's enough to persuade the body not to do what it's ready and eager to do: make babies.

Receptors in various organs are specifically attuned to receive the particular hormone that part of the body needs. Breasts, for example, are full of receptors for female hormones, the oestrogens. Prostates are full of receptors for male hormones, the androgens. I picture the receptors as something like a lock, and each kind of hormone is a key. The right key will open the lock and let the hormone signal through.

Scientists have recently discovered that a lot of musk chemicals happen to fit those locks. The lock can't tell the difference between the real key and the fake one. It lets the chemical in as if it were the real thing.

Artificial musks can interact with the hormone system in various ways. Some can mimic thyroid hormones.[16] Others mimic oestrogen.[17] This doesn't just happen with musks being studied in the lab: ordinary commercial products do it, too. One study looked at the fragrance in spray deodorants and found that seven out of the ten they analysed showed 'estrogenic activity'—that is, the body responded to the fragrance as if it were oestrogen.[18]

Some musks block oestrogens. In a 2010 study researchers found that one of the artificial musks could partially block the action of the body's own oestrogen: the more of the musk there was, the more it inhibited the oestrogen.[19] Some musks could block both oestrogen and progesterone, another important reproduction hormone. They did this at low doses, prompting the researchers to urge further study of the potential hormone-disrupting effects of these chemicals.[20]

Other musks interfere with the male sex hormones. Some musks can mimic androgens and others can block them.[21] Others, strangely, can do both blocking and mimicking actions, depending on the amount and time of delivery.[22]

—

Musks aren't the only fragrance ingredients that can fool the body into thinking they're its own hormones. Scientists have found that several of the compounds used in fragrance to block ultraviolet light also have hormone-like effects.[23] So does the universal blender-and-fixer found in virtually all fragrances, the phthalate DEP.[24] DEP doesn't bioaccumulate the way musks do, but many people get a fresh dose every day from fragranced products.

(Fragrance chemicals aren't the only sources of hormone-type chemicals. Other newly developed chemicals—among them, pesticides and plastic additives—can mimic hormones, or block them. It's not just the modern, artificial chemicals, either: some plants have chemicals—phytoestrogens—that mimic oestrogen. All these non-body sources of hormones are called xenoestrogens—foreign oestrogens. Their effects on human health are still being explored.)

—

One way and another, it became apparent that these fragrance ingredients could interfere with the body's signalling system. Some studies found that these interfering effects were weak—at least compared with the body's own hormone signals. And, for some of them, high doses were needed to get any effect. But other studies found relatively strong effects at relatively low doses.[25] Some found oestrogenic activity not just in the lab, but at levels that are found in the environment.[26] Further studies suggested the design of earlier tests might have meant the hormone effects had been underestimated.[27] Even studies finding a weak effect also concluded that there could be a cumulative effect.[28]

There can be strange effects with low and high doses of hormones. With regular poisons there'll be a conventional upward-curving graph: the more poison you're exposed to, the sicker you get. But with hormones the graph can be a U or upside-down-U shape—big effects can happen with either small or large doses. This is because the hormone system is a bio-feedback loop, constantly balancing out the amount of hormones to match the body's needs. As Professor Frederick vom Saal, a researcher at the University of Missouri who's spent years studying hormone-disrupting chemicals, puts it, 'Sex hormones cause opposite effects at high and low doses...At high doses, they turn off responses that they stimulate at low doses.'[29]

What began as an accidental finding by sludge researchers has, over the last few decades, become a substantial body of science. We know now that some of the chemicals in artificial fragrance have the remarkable ability to fool the body into thinking they're part of itself. The question is, could that be a problem?

CHAPTER 11

UNINTENDED
CONSEQUENCES

When lab animals were exposed to hormone-like chemicals, they didn't generally get sick, or suffer neurological damage, or get cancer. But some strange things were happening, and once again it was researchers working around wastewater plants who were the first to notice. A lot of gender confusion was going on among fish and mussels and other aquatic creatures living downstream from the sewage plants. They often had abnormalities in their reproductive organs that the scientists called 'imposex': one sex being imposed on the opposite sex. Males were 'feminised' and females were 'masculinised'. Sex ratios were skewed, so that there was abnormally more of one sex than the other. Distorted sex organs occurred in both sexes, including what the researchers called 'superfemales' whose reproductive organs were exaggerated to the point of deformity.[1]

Other scientists started to investigate the effects of those chemicals on mammals, rather than fish. In lab animals such as rats, a pattern began to emerge. Adult rats exposed to hormone mimics didn't have too many problems, but their offspring did. Male rats exposed to oestrogen-mimicking chemicals while in the womb often suffered certain kinds of abnormalities in their genitals. These included non-descending testicles, abnormal positioning of the urethra opening on the penis, and a reduced distance between anus and genitals.[2]

Researchers believe that this set of abnormalities—this syndrome—may point to an underlying pattern that has to do with hormones: specifically, not enough male hormone getting through to male rats during their development. As the scientists put it, the evidence suggests that 'the syndrome goes back to diminished androgen action in foetal life.'[3]

One reason for this conclusion is that they know descent of the testicles is triggered by androgens. If those male hormones haven't got through, then the testicles may not descend.[4] They also know the abnormal-urethra-opening disorder is critically connected to hormones that the foetus is exposed to.[5] And their studies suggest that abnormally reduced anogenital distance is associated with not getting enough male hormone during foetal life.[6] (Female rats have a shorter distance between anus and genitals than male ones—it's how you tell the difference between the boys and the girls when they're babies. This leads researchers to suggest that a shorter anogenital distance in a male rat may mean the rat is somewhat 'feminised'.)

What all this means is that, if the balance between male and female hormones is disrupted in the womb, the development of male plumbing can also be disrupted. Many researchers believe that the particular nature of this skewed development is an outward sign of what they call 'incomplete masculinisation'. The idea is that, at a vital stage of development, the foetus has received confusing messages from male and female hormones about what kind of genitals to make.

By the early years of the twenty-first century, there'd been so many studies that showed these kinds of problems that the hormone-mimicking chemicals had acquired a jokey nickname: 'gender-benders'. A lot was still unknown, but one thing was clear: these hormone-like chemicals weren't neutral substances. They were having profound effects—at least on mussels, fish, rats and little clumps of human cells in test tubes. But there was one big problem with these studies. They weren't about people.

—

A German study from 2001 was one of the early windows onto the effects of these chemicals on humans. When researchers in a clinic treating people with hormone disorders at the Heidelberg University Hospital thought to test their patients for artificial musks, they found that ninety per cent had at least two different kinds in their blood. The greater concentration of musk each woman had in her blood, the more severe her hormone disorder was. The researchers considered that the musks 'may act centrally as a disrupter of the (supra-) hypothalamic-ovarian axis...a reproductive toxicity and an endocrine effect of NMCs [nitro musks] in women cannot be ruled out.'[7] In other words, the musks were thought to have had an effect on the hormone system which may have been linked to the women's hormone and reproductive problems.

What happens when the human body, going about its ordinary hormonal business, gets a big extra hit of something that it thinks is oestrogen? What happens when something blocks testosterone from getting through? As with the rats, the worst things don't happen to the adults, but to babies in the womb.

Until it's born, a baby shares its mother's bloodstream. The placenta screens out some toxins from that blood to prevent them getting into the baby's body, but by no means all. If the mother is using a lot of fragranced products, some of those substances can be absorbed through her skin, lungs and digestive system, and travel in her blood into the baby's body. That body is going through a series of massive changes: in nine months it has to turn from a two-celled blob into a human being. A lot of that development is controlled by hormones. If the hormone signalling is disrupted, there's a risk the development will be, too.

A basic example is whether a baby is a boy or a girl. All foetuses start life as female. But, at a certain point in development, around half of them get a signal that makes them develop into boys. That signal is a shot of boy-making hormones, and the shot has to arrive during a short period of time called the 'male programming window'. For boy-making to happen properly, there's got to be exactly the right amount of hormone at exactly the right time.

The trouble is, some of those fragrance chemicals can block the male hormone. When that happens, the hormone doesn't get through, or not in the right amount or at the right time. It's even worse if the system is being flooded with female hormones at the same time. Less boy-making hormone getting through, plus lots of girl-making hormone: not surprisingly, the developing baby isn't sure what message it's getting. Should it be listening to the boy-hormone signal or the girl-hormone signal?

In rats, as we've seen, an oversupply of girl hormone can skew the development of the male sex organs in a specific way: the testicles not descending, the urethra opening in the wrong place, and a reduced anogenital distance.

You can't deliberately manipulate human embryos to see if the same thing happens in us. But, as it happens, a group of human embryos *was* manipulated in exactly this way. For some thirty years, starting around 1940, several million women around the world were given a drug called DES (diethylstilboestrol) in a bid to help them with fertility problems. DES was a synthetic oestrogen. That meant the children of those DES-treated mothers were getting a big dose of extra oestrogen as they developed in the womb. These children, and their own children, are giving researchers the chance to study an accidental, and sometimes tragic, experiment: what happens when you expose human embryos to an extra hit of oestrogen?

'DES sons'—the sons of women who took DES—show a greater-than-normal incidence of undescended testicles.[8] They're more likely to be born with an abnormally placed urethra opening.[9] A generation further on, 'DES grand-sons' also have an increased risk of suffering from the misplaced-urethra-opening disorder.[10] This means that the pattern of abnormalities in humans exposed to excess oestrogen shows similarities to the pattern seen in rats.

Working backwards from the observed pattern to a possible cause, researchers studied boys with reduced anogenital distance. Those boys, it was found, tended to have mothers who had high levels of oestrogen mimic in their blood. (The oestrogen mimic this particular study looked at was DEP, the phthalate found in fragrance, and this study looked at levels of the chemical that are found in the real world, not exaggerated experimental levels.)

The researchers summed up their findings like this: 'These data support the hypothesis that prenatal phthalate exposure at environmental levels can adversely affect male reproductive development in humans.'[11]

Another study found that boys who were born with lower androgen levels than normal—less of the virile male hormone—tended to have mothers with high levels of oestrogen mimic in their blood. It was as if the male hormone was being cancelled out by the female hormone. 'Our findings', the authors reported, 'are also in line with other recent human data showing incomplete virilization in infant boys exposed to phthalates prenatally.'[12]

These studies reinforced the idea that extra oestrogen does indeed have a gender-bending effect on developing male humans. Researchers call this effect 'testicular dysgenesis syndrome'—abnormal development of the genitals.[13] The problem with this isn't just the immediate, visible effects such as undescended testicles. Basic structures in the secret depths of male plumbing may have gone awry in the womb, setting up welcoming conditions for problems to occur decades later.

—

Prostate cancer is a hormonal disorder: it's about having an excess of male hormones in your prostate. The main treatments are designed to reduce all that androgen.

But where does the excess androgen come from? One

source, paradoxically, may be oestrogen. Several studies on rats have demonstrated that, if the prostate is exposed to extra oestrogen while it's developing, the risk of later prostate cancer can increase.[14]

Researchers think this may be because, as the male develops in the womb, his prostate makes a certain number of receptors for androgen. This will enable his prostate to make sperm and do other manly things later on. But if his prostate is flooded with a lot of extra oestrogen while it's still developing, it gets alarmed. Too much female hormone! Quick, make some more male hormone receptors! Perhaps, to compensate for all that oestrogen, his prostate may make more androgen receptors.

One study (of mice, but researchers think there may be a similar effect in humans) found that a fifty-per-cent increase in oestrogen during foetal life led to a doubling of androgen receptors. This was a permanent effect— as adults, these mice had more androgen receptors and bigger prostates.[15]

Having a lot of androgen receptors means that when the male becomes an adult, his prostate is saturated in androgen. Prostate cancer loves androgen: the more androgen receptors you have, the more welcoming the conditions are for cancer.

That extra hit of oestrogen (or an oestrogen mimic) the baby receives in the womb, triggering the creation of more androgen receptors, could have come from many sources. There'd have been some from the pesticides on the vegetables his mother ate. There'd have been more from the plastics in which she stored and microwaved her food. And there'd have been another dose from the fragranced products she used: from the good-smelling musks, the ultraviolet-light blockers and the phthalate DEP.

Did any of that extra oestrogen have anything to do with the later cancer? In any particular case, no one can say—there are too many variables over an individual's lifetime to be able to point the finger at one 'cause'. But more and more threads of research are connecting oestrogen during development and prostate cancer, and a plausible mechanism has been proposed to explain that link. Given all that, it seems sensible for pregnant women to take the precautionary path and avoid as many sources of gender-benders as they can—including artificial fragrances.

—

Prostate cancer is the most commonly diagnosed cancer among men.[16] For women, it's breast cancer. Over a lifetime, a woman's risk of being diagnosed with breast cancer is around one in eight.[17]

Not all breast cancers have a hormonal basis, but more than two-thirds do. They're the kind called 'oestrogen-receptor-positive'.[18] This means that they rely on oestrogen for growth. One of the main chemotherapy drugs for oestrogen-receptor breast cancer, Tamoxifen, is a chemical that blocks oestrogen receptors.

Many studies point towards the conclusion reached by a committee of European Commission researchers about breast-cancer risk: 'the more estrogen reaches the sensitive structures in the breast during [a woman's] lifetime, the higher the overall risk.'[19]

The unintended experiment with DES—the oestrogen drug given to women to enhance their fertility—is a valuable source of information for female cancers. DES mothers—the adult women given the drug—were found to have a higher chance of developing breast cancer than other women. DES daughters—exposed while they were in the womb—also have an increased breast-cancer risk.[20]

DES daughters also have a much greater (by about forty times) chance of developing a rare cancer, adenocarcinoma. They also have a greater-than-normal chance of having abnormal pre-cancer cells in their cervix or vagina. They're more likely to have a cervix or uterus of unusual shape and structure, and have tended to have various problems with getting pregnant and carrying a baby successfully to term.[21] (These unintended consequences weren't discovered until some thirty years after DES was first prescribed. After 1971 the drug was no longer given to pregnant women.)

Hormone Replacement Therapy is another of those well-meaning experiments on humans. HRT is a therapy designed to reduce the symptoms of menopause, when women's hormonal levels change. HRT adds an extra dose of oestrogen and other hormones. As with DES, adding a lot of extra female hormones turned out not to be such a good idea.

A big study started in 1991 gave HRT to one group of women and not to another. The effects were so dramatic that the study had to be stopped early for ethical reasons: the women getting HRT were developing many more invasive breast cancers than the others.[22] The Million Women Study in the UK, a similar experiment with HRT, confirmed these results. The *Lancet*'s conclusion about that study is unambiguous: 'Current use of HRT is associated with an increased risk of incident and fatal breast cancer.'[23] When the results of these studies became public, many women stopped HRT. (In the few years following, there was a temporary downward blip in the increase of new breast cancers, especially among older women, who were most likely to have been using HRT.[24])

An extensive review of all the scientific studies about hormone-disrupting chemicals concludes that recent studies 'have added weight to the idea that any estrogen, including xenoestrogens, might contribute to breast cancer risks'.[25]

—

As with prostate-cancer risk, how oestrogen may increase breast-cancer risk is still not fully understood, but there are theories. Some researchers think it has to do with excess oestrogen exposure in the womb.

As a girl baby grows in the womb, at a certain point she gets a signal from the hormone oestrogen to start making structures called end buds.[26] During puberty, these blossom into adult breast structures. The end buds are where the oestrogen receptors in the fully developed breast are located.

In a pregnant woman exposed to a lot of oestrogen, or oestrogen mimics, her unborn daughter will be getting more oestrogen than she would otherwise. It's oestrogen that triggers the formation of these end buds, and it's end buds that are involved in breast cancer. The finding of one study was that 'The majority of breast cancers in Western women derive from end buds, where the cells that contain estrogen receptors and are most responsive to estrogen during breast development are located.'[27]

You can't draw a straight line between fragrance and breast cancer, of course. But the most common and most steadily increasing kind of breast cancer is linked with oestrogen. At the same time chemicals in our environment, including the ones in fragrance, are giving us an excess of oestrogen and oestrogen mimics. There may be no connection between these two facts. Still, they suggest it makes sense to avoid as much environmental oestrogen as you can.

—

Prostate and breast cancers aren't the only diseases that are linked to the sex hormones. In the case of male disorders, studies have also shown links between oestrogen and testicular cancer. One found that 'Exposure of the mother to exogenous estrogen during pregnancy created a significant risk in the son.'[28] A later study found biological pathways that could explain this link.[29] There's also an association between oestrogen and poor sperm quality, which leads to fertility problems.[30] For women, oestrogen-related disorders in addition to breast cancer include puberty happening too early, difficulties with conceiving and bringing a baby successfully to term, polycystic ovary syndrome, endometriosis, uterine fibroids, ovarian cancer, and the kind of endometrial cancer that's hormone-dependent.[31]

All these disorders, male and female, have increased over the last few decades. A detailed survey of the available data concluded that, when it comes to prostate cancer, 'All European countries (except in the high incidence countries The Netherlands and Austria) have in recent years experienced dramatically increasing incidence trends.'[32] Testicular cancer has increased in young men in a way the researchers describe as 'an unexplained epidemic'.[33] Sperm quality has declined, along with fertility, especially in younger men: 'more than 1 in 6 have an abnormally low sperm count.'[34] The incidence of undescended-testicles disorder has increased.[35] So has the abnormal-urethra-opening disorder. In a study from Western Australia the prevalence of this disorder increased by two per cent per annum over two decades, so that it now affects one in 231 births.[36]

Even allowing for greater awareness and testing, the increase in breast cancer is especially marked. In 1982 in Australia there were 5,368 new cases of breast cancer diagnosed. In 2011 there were 14,568. The age-standardised incidence rate over the same period increased from forty-four to sixty cases per hundred thousand people.[37] Of that growth in numbers, it's the oestrogen-receptor-positive breast cancers that have increased most. One American study examining the relative increase in hormonal and non-hormonal breast cancers concluded that, over the period of the study, 'the proportion of tumors that are hormone-receptor-positive rose as the proportion of hormone-receptor-negative tumors declined...the overall rise in breast cancer incidence rates in the United States seems to be primarily a result of the increase in the incidence of hormone-receptor-positive tumors.'[38]

Other female disorders associated with oestrogen have also become more common over the last few decades: precocious puberty,[39] ovarian cancer and one type of endometrial cancer, the one that's dependent on oestrogen.[40]

No one would claim that the increase in all these disorders is single-handedly down to the hormone disrupters in fragrance. Fragrance is only one source of hormone disrupters, and in any case there hasn't yet been enough research to know exactly why hormonal disorders are on the rise. But scientists are concerned. They consider that the speed of the increase of these problems, and in some cases their geographical pattern, strongly suggests that environmental factors play a key role. The hormone-disrupting chemicals in fragrance are a ubiquitous and increasing part of that environment.

—

Ten years and many studies after the Heidelberg researchers noticed a correlation between musks and hormone disorders, the European Commission funded a major review by a panel of independent scientists that looked at all the studies into hormone disrupters produced up to that point. The result is a document, published in 2011, which represents, as the title says, the current 'State of the Science of Endocrine Disrupting Chemicals'.[41] If you were to try to cram those five hundred dense pages into one sentence, it might be something like: There isn't enough data to know for sure, but the available data, as well as our understanding of how hormones work, suggest that hormone-disrupting chemicals are likely to be doing us damage.

In the US there's also concern over consumer safety. Around 2009 a committee of scientists funded by the Consumer Products Safety Commission was given the job of assessing the hormone disrupter DEP, used in fragrance, for its health effects. The committee scrutinised every study that had been done on DEP. Their report, published in 2014, concluded, 'Exposure to DEP can induce reproductive or (nonreproductive) developmental effects in humans...[the studies] suggest that harmful effects in humans have occurred at current exposure levels. There is, therefore, an urgent need to implement measures that lead to reductions in exposures, particularly for pregnant women and women of childbearing age.'[42]

—

Is it possible, though, that the scientists are being alarmist? Should we be worried? After all, the doses of the hormone chemicals are small and their effects are relatively weak.

It depends on how small is safe, and how weak is safe, and it depends on what happens over time. Many exposures to many weak oestrogens, every day of the year, might be adding up to a dose that will make a difference to our health. And, as we've seen, with hormones even a tiny amount can have a big effect. How small is safe, how weak is safe and what the long-term effects might be, are questions no one yet has the answers to.

The other thing we don't know is how much of these chemicals we're being exposed to. As is the case with carcinogens, we can't know whether they're in our shampoo or scent because musks, being a 'parfum', don't have to be declared on the labels of the products they're in. And one of the biggest sources of musks, laundry detergents, don't even have to use the trade-secret hidey-hole, because clean-ing products aren't required to list their ingredients.

—

Research into these new chemicals is as new as research into smoking was fifty years ago. But the people who sell fragranced products are watching the science about hormone disruptors, and some of them are acting, even though no law yet says they have to.

As far back as 2005, the Body Shop stated that they were phasing out the older kinds of the artificial musks about which there were health concerns (nitro musks and polycyclic musks) and replacing them with the newer kinds (macrocyclic musks). They also undertook to phase out the phthalate DEP by 2008. They said, 'we have chosen to take a precautionary approach...even though these ingredi-ents are legal and considered safe for use by our industry and its regulators.' By 2010 the company had phased out the phthalates. In the case of the musks, their website states that, for animal-welfare reasons, they now only use synthetic musks rather than the animal-based ones. They don't, however, specify which synthetic musks they use.[43]

As we saw, in 2011 Johnson & Johnson undertook to remove two carcinogens from its baby products. At the same time it also undertook to remove phthalates from all the products they made for babies, and by 2015 they announced that they'd done so.[44]

In doing this, these companies were responding to the science, and to growing public awareness. Consumers expect chicken and beef that are 'hormone-free', and we're glad that our roll of cling wrap or water bottle declares itself to be free of the oestrogen mimic BPA. But, given the amount of hormone disrupters we may be getting from laundry detergent—something that's up against our skin all day and all night—and all our fragranced personal-care products, we might be ignoring bigger sources of risk. And after all, to put it bluntly: the more hormone disrupters you can avoid, the less your hormones will be disrupted.

CHAPTER 12

SHARING THE AIR

For fragrance lovers, their perfume is a beautiful thing that brings great pleasure, as intricate and interesting as a fine wine. It's an enrichment of life, like wonderful music. Finding the right perfume is another way of feeling good about yourself. A perfume makes a statement about you, the way your choice of clothes does. It's part of the subtle signalling that's one of the delights of being a social animal. In a crowded world, making a personal bubble of your own smell is a way of marking out a bit of territory that belongs to you.

But for people who get sick from fragrance, it's a tyrant that more or less runs (and ruins) their lives. They're likely to suffer every time they leave the house. Going shopping, getting on a bus or train, eating in a restaurant, going to the movies, going to work, going to their uncle's funeral—they'll probably get a headache or breathing problems when they do any of those things, because in all those places there's sure to be at least one source of artificial fragrance. If it's not the strong perfume or cologne someone's wearing, it'll be the air freshener or the scented candles, or the fragrance in the laundry aisle in the supermarket or the incense at the yoga class (it's been a long time since incense was made from real frankincense). Working and studying will be difficult—it's hard to think well when you've got a headache or you're wheezing. Socialising will be a toss-up between enjoying friends' company and getting symptoms, so sufferers' lives will shrink to a few safe places and a few safe companions. Rubbing salt in the wound will be the fact that many other people won't believe the problems are caused by fragrance.

People have the right to breathe air that doesn't make them sick. Do people also have the right to smell the way they want to? People on each side would say their rights are legitimate. In Australia, no compromise has been worked out yet between these contradictory rights. Even the existence of the problem is a new idea. But, as Susan McBride in that office in Detroit demonstrated, this is a dilemma that's not going to go away. More and more people are getting sick from fragrance, and somehow those competing rights have to be accommodated.

There's a school of thought that draws a parallel between second-hand scent and second-hand smoking. In both cases, the problem is a fundamental biological one: even if you don't like what's in the air, you can't choose to stop breathing.

Until thirty years ago we dealt with the conflicting rights of smokers and non-smokers in a very simple way: by denying the rights of non-smokers. By afternoon, offices and school staffrooms everywhere were blue with smoke. People smoked in the movies, in restaurants, even on planes. If you didn't want to breathe other people's smoke, too bad.

Two things changed all that. One was the medical evidence on the dangers of second-hand smoke. By the 1980s there was no longer any room for argument: breathing other people's smoke was nearly as bad for you as doing the smoking yourself. The other big change was legislation in many countries that protected people from dangers in their workplaces. These days, employers are obliged to provide a safe workplace, and this includes providing air that's safe to breathe.

Once second-hand smoke was shown beyond argument to be a hazard, employers were obliged to protect their workers from it. The only way to do that was to ban smoking in every kind of workplace, and the only way to enforce the ban was to cement it into place with legislation.

But, even in countries where workplace smoking is banned, smoking isn't illegal. Smokers still have the right to light up. They're free to damage their own health—they're just not allowed to damage other people's. Like all compromises, it's not a perfect solution for either side. Smokers often have to huddle in doorways if they want a cigarette, or wait till they get home. And if your neighbours smoke on their balcony and the smoke wafts into your bedroom, you have to put up with it. But it's a solution that steers some sort of path between conflicting rights.

It seems impossible to imagine anything like that happening with fragrance. But back in 1960, the idea of banning smoking would have seemed absurd. Only a crackpot would have predicted that smoking would one day be banned in public places. Yet here we are today, in a world where crackpot is the new normal.

—

Susan McBride's win had an almost immediate flow-on effect in the US. Within a week, the City of Detroit brought in a fragrance-free policy in all its workplaces. Others soon followed. The Centers for Disease Control and Prevention includes fragranced products on its list of 'Chemical Contaminant Sources' and has made all its workplaces fragrance-free—fifteen thousand employees, in dozens of offices across the US.[1] The US Census Bureau, with thousands of employees, issued a fragrance-free policy the same year. Many other organisations, big and small, public and private, followed suit: workplaces, schools, colleges and hospitals all over the country. Even some theatres and churches now set aside 'fragrance-free zones'.[2]

Fragrance-free policies receive a strong majority of support: in a recent study, fifty-three per cent of people would support a fragrance-free policy in their workplace, compared to less than twenty per cent who would not. That is, 2.7 times more people would vote yes for such a policy than would vote no.[3]

When the Americans brought in these policies, they didn't have to reinvent the wheel—their neighbours in Canada had already shown how it might be done. As far back as 1991, nurses at the Queen Elizabeth II Hospital in Halifax, Nova Scotia, had brought about a fragrance-free workplace, for their patients' sake and their own.[4] Many other places in Canada did likewise: hospitals, workplaces, schools, universities, even some concert venues, are fragrance-free. Vancouver International Airport has a 'Fragrance Free Route / Itinéraire sans parfum' through its duty-free shops.[5]

Canadian laws about disability specifically cover 'environmental sensitivity', which includes getting sick from fragrance. They make no bones about the reality of the problem and the employer's obligations to address it:

> Individuals with environmental sensitivities experi-
> ence a variety of adverse reactions to environment

agents at concentrations well below those that might affect 'the average person'. This medical condition is a disability and those living with environmental sensitivities are entitled to the protection of the *Canadian Human Rights Act*, which prohibits discrimination on the basis of disability...Like others with a disability, those with environmental sensitivities are required by law to be accommodated.[6]

Fragrance-free workplaces in North America have worked out policies worded like this, from the Centers for Disease Control:

Scented or fragranced products are prohibited at all times in all interior space owned, rented, or leased by CDC. This includes the use of:
- Incense, candles or reed diffusers
- Fragrance-emitting devices of any kind...
- Potpourri
- Plug-in or spray air fresheners
- Urinal or toilet blocks
- Other fragranced deodorizer/re-odorizer products...

In addition, CDC encourages employees to be as fragrance-free as possible when they arrive in the workplace. Fragrance is not appropriate for a professional work environment, and the use of some products with fragrance may be detrimental to the health of workers with chemical sensitivities, allergies, asthma, and chronic headaches/migraines.[7]

Fragrance-free workplaces in the US have signs up like this one, from the Massachusetts Nurses' Association: 'Welcome. This is a Fragrance Free Health Care Environment. For the health and comfort of all who use this facility, kindly avoid using fragrance.'[8]

Many public events in the US—weddings and conferences among them—now issue a request along these lines: 'The conference will be a fragrance-free event. In order to help every participant enjoy the conference, please be mindful what you wear. This includes scents in shampoos, hair styling products, laundry detergent, deodorants, essential oils and perfumes.'

No doubt there are people unhappy with these policies and people who ignore them. Perhaps some of those fragrance-free workplaces are still full of fragrance. But a publicly announced policy gives everyone a way of talking through the options for any particular workplace. The possibility of an expensive court case gives the employers an incentive to find a way to make the policy work for everyone.

—

Safe Work Australia is our national body to ensure workplace safety. It says, 'a person conducting a business or undertaking must take steps to eliminate or, if the risk cannot be eliminated, minimise the risk of exposure to workers from hazardous chemicals and the risk of exposure to airborne contaminants.'[9] 'Health Hazards' in a chemical are

> properties of a chemical that have the potential to cause adverse health effects. Exposure usually occurs through inhalation, skin contact or ingestion. Adverse health effects can be acute (short term) or chronic (long term). Typical acute effects include headaches, nausea or vomiting and skin corrosion, while chronic health effects include asthma, dermatitis, nerve damage or cancer.[10]

As we've seen, many fragrance chemicals can have these effects.

No one in Australia has won a landmark case the way Susan McBride won that case in Detroit, but we have the legal anti-discrimination framework that could make this possible. The Australian Human Rights Commission recognises that sensitivity to chemicals can be a barrier to access, pointing out that 'A growing number of people report being affected by sensitivity to chemicals used in the building, maintenance and operation of premises. This can mean that premises are effectively inaccessible to people with chemical sensitivity.' The ability to access premises and services is a basic right upheld by our anti-discrimination laws.

The Human Rights Commission is clear on the obligation to address issues of accessibility: 'If you have identified some access barriers or gaps in your services they need to

be fixed as soon as possible in order to avoid continu-
ing discrimination.'[11] This means that if, for example, an
asthmatic was unable to enter a rest room because an air
freshener was in use there, the owner of the premises could
be liable under these laws.

~

'Jane Smith', the Australian public servant who got sick
from fragrance, lost her case, but another plaintiff—
perhaps one with a good lawyer—might win a similar
case. That will change everything.

Yet the big guns of the law mightn't be necessary. Just
having a fragrance-free policy (no matter how voluntary, no
matter how unenforceable) has a powerful effect. Its very
existence sends the basic—and often new—message: there
are people who get sick from other people's fragrance. The
issue is framed in a constructive way: it's not a matter of
'problem people'. Workplace fragrance becomes a routine
matter, another thing that a good manager has to work
through with staff.

The WorkplaceOHS website (a federal-government
service) runs an advice forum. In March 2008 it received
this question from a workplace occupational-health-and-
safety officer:

> Can we enforce a 'no personal spray' policy? We
> have asthmatics in the workplace who find spraying
> of toilet deodoriser and perfume in our workplace

and the toilets, triggers asthma. Although we have sent out messages several times asking all employees not to use sprays in the workplace, this request is ignored by some, as they believe the asthmatics are just complainers. Is there some way to enforce the no spray alert in the workplace, to please everyone, and to eliminate this health issue in the work environment?

WorkplaceOHS answered:

Perfumes or any scents can adversely affect workers' health, causing headaches, nausea, dizziness, upper respiratory symptoms, skin irritation and difficulty with concentration. Further, certain odours, even in small amounts, can trigger allergic and asthmatic attacks in workers with the condition. So, it is in employers' best interests to address the issue of perfumes and personal sprays in the workplace. To reduce the risk to the asthmatics in your workplace, consultation should be undertaken with your safety representative to develop an air-contaminants or scent-free policy. The developed policy can then be communicated to all employees and alerts posted in the amenities outlining and re-enforcing the policy. This approach will also make employees aware that there are asthmatics within the workplace.[12]

Even without a push from the OH&S people, more and more organisations are encouraging low-scent workplaces. Several government offices in Canberra now have posters up asking people to leave the scent at home for the sake of their fellow workers. Choirs routinely insist on their singers being fragrance-free, as it's hard to sing well when your throat is irritated by perfume. And, in a growing number of workplaces, people agree among themselves not to use heavy scents, out of consideration for others. These days many people can see that filling the air with a strong perfume is a kind of invasion of everyone else's personal space.

—

Not everything in our lives is the subject of a regulation or a policy. As social animals, we know that the things we do have an effect on other people, and we've evolved ways of behaving that we might call etiquette. We cover our noses when we sneeze and cover our mouths when we cough. We eat with our mouths closed. We don't queue-jump, we listen to our music through headphones and we try not to fart in public.

We don't need rules and regulations to tell us any of this. As adults in a civilised society where we all live at close quarters, we pay each other these courtesies. We're aware of other people's personal space, and hope they'll be aware of ours. There's no law that says you can't play the bagpipes on the bus, but everyone's glad if you don't.

CHAPTER 13

THE OPPOSITE OF FRAGRANCE

As I researched the effects of fragrance I came across the website for the Canadian orchestra Symphony Nova Scotia. Among the FAQs on their site they say: 'Since several of our patrons have severe scent allergies, we ask that you please leave the cologne, perfume, hairspray, and deodorant at home!'[1]

So when a man from the Sydney Opera House rang me to get feedback on *Don Giovanni*, I was emboldened by that example.

The music was wonderful, I said. The only problem was a woman near me—her fragrance gave me a terrible headache.

I listened to a long silence on the end of the phone. I pictured him rolling his eyes: *Uh-oh, we've got a loony here.*

Really, he said at last. You mean perfume?

Yes, I said. A lot of people are made sick by other people's perfumes.

Oh, he said. I'd never heard that.

Yes, I said in my mildest way. It's not talked about much, but it's quite common. In fact, there are orchestras overseas that ask their patrons not to wear fragrance.

Oh really, he said. I could hear his disbelief.

Symphony Nova Scotia, for example, I said. In Canada. Google them.

They ask people not to wear fragrance, he said. It wasn't a question. It was more the flat tone you'd use with the wild-eyed person on the street who's claiming to have been abducted by aliens.

Yes, I said, it's for the patrons, but the musicians, too.

Well, he said. I'll certainly make a note of that, Mrs Grenfell. Thank you so much for your feedback—we certainly appreciate it.

In the moment after we both hung up I could imagine the laughter ringing around his office. This nutter on the phone, he'd be saying, you won't believe it, something about getting sick from perfume!

It's uncomfortable, making a laughing stock of yourself. Well, I thought, I won't subject myself to that again. But then I pictured him for the rest of the day, making a funny story out of it, sharing the joke with his mates in the pub after work, with his partner when he got home.

Statistically, there's a good chance that at least one of those people would have a problem with fragrance. At that very moment one of them might be saying, 'Actually, now you mention it...'

~

Planet Fragrance is a new thing. Anyone born before about 1960 grew up in a world that was chemically completely unlike today's. The only fragrances came from flowers, or from those rarely used tiny bottles on dressing tables. There were no fragranced detergents or laundry powders. Toilet paper smelled of nothing worse than paper. If you wanted to freshen the air, you opened the window. If you wanted your toilet to smell clean, you cleaned it. You might have been given a cake of rose-geranium soap, or lily-of-the-valley talcum powder, for your birthday or Christmas—but something fragranced was special.

That was because fragrance used to cost a lot. When Joy was first marketed, in 1933, a bottle cost a week's wages. These days, no one needs to spend a week's wages on perfume, even top-of-the-line stuff. Synthetic ingredients have turned the economics of fragrance inside out. The fancy perfume bottle can cost more to make than the liquid inside it, and the marketing costs much more than either.

But even when fragrance got cheaper, it never lost

its glamour—marketing sees to that. The combination of cheapness and glamour means that no one keeps perfume for special occasions the way Mum used to. And fragrance isn't just in those little bottles now. Saturation advertising has normalised flowery and lemony and piny smells in everything from toilet paper to garbage bags, to the point where fragrance is taken for granted as an essential part of life.

Luca Turin, the aftershave-with-carcinogens man, considers perfume to be an art form. He believes that smelling a great perfume is as glorious an experience as hearing a symphony. His life has been devoted to all things fragrance. But even he says that there's far too much fragrance around these days.

> What has changed, and not for the better, is the shift from symphony to jingle...This is due to a combination of factors: (1) too many launches, more than five hundred a year...(2) the profitability of aroma-chemicals [synthetic aromas] and the cheapening of formulas, which means the big firms tolerate expensive naturals [essential oils] only if nothing else will do, and (3) the necessity for a fragrance to shout even louder to make itself heard.[2]

Passionate as he is about perfume, Turin would like to remind us that we won't be thrown off the bus or expelled from the opera if we've failed to squirt ourselves with perfume, and that there are times when being fragranced isn't appropriate. 'If you are going to a film or a performance, during which other ticket holders will be forced to sit next to you for hours, have the decency either to wear nothing or to wear something that leaves some air in the room.'[3] Of one popular and especially pungent perfume he pleads: 'Please never, ever wear it to dinner.'[4]

It's true that, by wearing a strong perfume, a woman can draw 'a whole new level of attention' to herself. She can walk through the office or a shop or along the street, leaving a slipstream of overpowering scent, pleased at making heads turn. She likes the smell, and she's probably experiencing olfactory fatigue, so she'll never guess that behind her back people are exchanging a look that says: 'Phew! What a pong!'

The thought of being without fragrance horrifies people who are used to it. Oh, but what about BO, they say. When you think about it, though, the opposite of fragrance doesn't have to be stink. There are plenty of good fragrance-free deodorants around. Within hours of the fragrance-free policy being announced at the City of Detroit offices, a clever deodorant company was giving out complimentary samples of their fragrance-free product to the employees there. Sales boomed.

That's the thing: you don't have to go without. Fragrance-free products are easy to buy these days. A quick survey of my local supermarket revealed that it now stocks four brands of fragrance-free laundry powder, plus fragrance-free soap, sunblock, bug repellent, dishwashing liquid, dishwasher tablets, shampoo, conditioner and other products. As I was writing this chapter, saturation advertising appeared on bus shelters all around Sydney for a new lipstick—the selling point, in enormous letters, was '100% fragrance-free'.

Some of these products are made with essential oils—they have 'no synthetic fragrance' rather than strictly being 'fragrance-free'. Essential oils can have some of the problems of synthetic fragrances, for the reasons we've seen. Still, they're better than those synthetics. If a manufacturer has gone to the trouble and expense of using essential oils, that will be a selling point, so every ingredient will be named on the label (rather than hidden behind the 'parfum' screen). You'll be able to see exactly what you're putting on your skin and breathing into your lungs. The other advantage is that essential oils are expensive, so manufacturers are likely to use the minimum amount to get a nice smell.

'Fragrance-free' products don't smell of anything other than whatever they're made of. That means they tend to be made with better raw materials. Most conventional personal and cleaning products are based on some version of mineral oil—refined petroleum—because that's the cheapest basis for them. (In spite of the hype, many expensive face creams are based on mineral oil, just like the cheaper versions.) But mineral oil doesn't smell very nice, so the product has to be fragranced. Products that are genuinely fragrance-free are based on ingredients that don't smell unpleasant: coconut oil or sweet almond oil, for instance.

Confusingly, 'unscented' isn't the same as 'fragrance-free'. An 'unscented' product has had an odour-neutralising chemical added to it. These odour-neutralising products are chemical inventions that lock on to various kinds of odour molecules and trap them. A better word for these products would be 'de-scented'.

Don't be fooled by greenwash. Manufacturers have spotted the market for clean and green products, and are experts at misleading. A product might tell you that it contains 'pure organic essential oils' or 'botanicals', and it might contain a drop or two, but it might also contain the usual cocktail of synthetic ones as well. The label may shout words like 'eco', 'natural', 'green', and 'earth', but this might mean only that the box is made from recycled paper. Something might claim to be 'lemon-fresh' and reinforce the impression with a picture of a lemon on the package, but it's virtually certain that no actual lemon had anything to do with its manufacture.

There's only one way to avoid synthetic chemicals—look for the f-word in the list of ingredients. No other claim can be believed. All those appealingly green words have no legal meaning whatsoever.

Alternative products are sometimes more expensive. Essential oil of jasmine is more expensive than synthetic jasmine, and sweet almond oil costs more than mineral oil. But not all alternative products are expensive—market pressure means that the price of these safer products is coming down all the time, so the price difference is less significant than it used to be.

Even when the alternative is more expensive, it's often not by much. The cheapest conventional laundry powder in my local supermarket costs $4.25 a kilo. The cheapest alternative laundry powder—with 'no synthetic fragrances'—costs $5.60 a kilo. That's $1.35 difference. Say you get about thirty loads of washing from a box, which is what the manufacturers promise. Between the mainstream detergent and the alternative one, that's a difference of under five cents per load.

If you can only afford one alternative product, spending a little extra on laundry powder is one of the best ways to make life safer for you and your family. This is because fragranced fabric is up against the skin more or less twenty-four hours a day, in conditions of heat and moisture. That makes the perfect environment for the maximum absorption of fragrance chemicals. A lot of that fragrance takes the form of the gender-bending musks that are particularly dangerous for children and the unborn.

Some fragrance-free products are actually cheaper than the fragranced version. The fragrance-free dishwashing liquid in my local supermarket, for instance, is cheaper by quite a margin than the fragranced kinds. Often, it's the cheaper, no-frills version of a product that's fragrance-free: toilet paper, tissues, nappies and candles are just a few. They'll do the job just as well, and reduce the load on both consumers and the environment.

—

Aromatherapy has a lot to answer for: there's a vague assumption that any kind of scent in the air must be good for you. This is one aspect of the current fashion for room fragrances: air fresheners, reed diffusers, scented candles, potpourri and incense.

The other reason for all those scented premises is the enormous profit to be made from supplying 'ambient scent-marketing'—that is, fragranced air in shops, casinos and hotels, sometimes created by scented candles or air fresheners, but increasingly pumped in through elaborate, high-tech air-circulating systems. The fragrance industry estimates this business to be worth around $300 million a year. It's based on the idea that people feel better and therefore spend more in businesses that are fragranced. This idea comes from research purportedly done by International Flavors and Fragrances (a fragrance-industry organisation), which claims that customers in ambient-scented stores spend twenty to thirty per cent more time shopping in them than in non-scented ones. However, when pressed by the media to share the actual research this claim is based on, International Flavors and Fragrances 'declined to provide [it,] on account of contractual agreements'.[5]

It's a tribute to the marketing subtlety of the fragrance industry that this claim has taken on the status of accepted wisdom—it was repeated to me, as fact, many times in my discussions with people about fragrance. If it's true, let's see the evidence. Given the gigantic profits in ambient scents—a whole new market for the fragrance industry— you don't have to be a cynic to wonder whether that 'research' really exists.

But research that shows the opposite is freely available. The 2016 study by Anne Steinemann quoted earlier found that thirty-five per cent of people experience health problems when exposed to fragrance. Not surprisingly, then, the same study found that twenty per cent said that, if they entered a fragranced shop or other business, they would leave it as quickly as possible. Fifty-five per cent said they would choose a hotel without fragranced air over a fragranced one. A study published in the journal *Chemical Senses* found that, contrary to expectations, 'ambient odors did not increase sales' in the outlets studied.[6]

For people who are affected by fragrance, scents in the air are as effective as a barbed-wire fence. Any shop, hotel, B&B, cafe, yoga studio or cab that uses any of those products won't get our business.

Fragrance sufferers often don't say anything—it's easier to go to the next shop or hotel or cab. Internet shopping is a godsend. Many of us now peer through the window of a fragranced shop, read the label on the product we like the look of, and go home and buy it online. I don't like doing that—if we want to have shops, we have to spend our money in them—but my spirits fail at the thought of yet another headache.

—

All over the world millions of people use fragrance every day, and for most of them it doesn't cause headaches or breathing problems. I understand how hard it is for those lucky folk to believe that many of us get sick from something that brings them such pleasure. But I like to think that, once they're aware, perfume fans will be prepared to think twice, recognising that a small sacrifice by one person can mean life-changing relief for others. Believe me, we'll be grateful!

As for the damage fragrance may do at a molecular, long-term level to ourselves or our unborn children—well, the people who make fragrance are confident their regulations protect us against that possibility. The evidence in this book, though, suggests their confidence could be misplaced. It is, after all, based on self-interest.

Still, it's true that our daily lives are full of risk. We know about the dangers of the BPA in our plastic water bottles, the pesticides on our broccoli and the hormones in our chicken wings. We've got sick of hearing about all the things that can give us cancer. If we turn away from using fragranced products, how many cancer or hormone problems won't we have? What difference will it make to the health of our unborn children? We'll never know. Unfortunately, you can't run your life twice, to see which of the choices you made were the ones that damaged you or your children.

We also can't know what the chances are that our house will burn down, yet most of us take out insurance to cover that possibility. We don't do that because we think it's likely to happen. We do it because of what's at stake. The odds may be small, but the consequences would be devastating.

One way of looking at a fragrance-free choice is as a kind of risk management. Our lives are full of risks we don't have any control over. But fragrance isn't one of them. It's not hard to make the choice to spend our days free of artificial fragrance, and discover that civilised life is perfectly possible without it.

—

Near the end of the promotion tour that made me realise how fragrance had changed my life, I was the speaker at a literary lunch. At the top table, toying with my chicken and wishing I didn't have a headache again, I sat next to the young man who was going to introduce me.

I owe you one, he said.

Oh, I said. How's that?

He told me the story: he'd started getting headaches. Weirdly, they'd begun the day after he'd come back from a trip to New Zealand. Then he was asked to introduce me at this literary lunch, and he went to my website for background information. He did a thorough job, and found the little piece I'd written all those years ago: 'Thanks for Not Wearing Fragrance'.

The light went on, Kate, he said. What I realised was, at the Auckland duty-free on the way home, I'd bought a bottle of cologne. I'd started using it the next day. When I read your piece, I joined the dots.

This book is for everyone like him.

NOTES

CHAPTER 1: PLANET FRAGRANCE

1. Pain, C., 'When Others Abhor the Fragrance You Adore', ABC Health & Wellbeing, 13 January 2015, abc.net.au/health/features/stories/2015/01/13/4160960.htm
2. Bridges, B., 'Fragrance: Emerging Health and Environmental Concerns', *Flavour and Fragrance Journal* 17, 5, 361–371, 2002.

CHAPTER 2: HOW MANY OF US ARE OUT THERE?

1. Rosen Law Office, 'A Fragrance-free Workplace', rosenlawoffice.com/a-fragrance-free-workplace
2. Towell, N., 'The Overpowering Scent of the Public Service', *Canberra Times*, 12 September 2014, canberratimes.com.au/national/public-service/the-overpowering-scent-of-the-public-service-20140912-10fmbu.html and tribunal findings at austlii.edu.au/cgi-bin/sinodisp/au/cases/cth/aat/2014/658.html
3. Discrimination Tribunal of the ACT, *Lewin v ACT Health & Community Care Service 2002*, austlii.edu.au/au/cases/act/ACTDT/2002/2.html
4. World Health Organization, 'How Common Are Headaches?', who.int/features/qa/25/en
5. Le, H., *et al.*, 'Increase in Self-reported Migraine Prevalence in the Danish Adult Population', *BMJ Open* 2, 4, 2012.
6. Stang, P. E. *et al.*, 'Incidence of Migraine Headache, a Population-based Study in Olmsted County, Minnesota', *Neurology* 42, 9, 1657–62, 1992.
7. World Health Organization, 'How Common Are Headaches?' and Steiner, T. *et al.*, 'Migraine: The Seventh Disabler', *Journal of Headache and Pain* 14, 1, 2013.

8. Migraine Research Foundation, 'Migraine Facts', migraine researchfoundation.org/about-migraine/migraine-facts

9. Spierings, E. *et al.*, 'Precipitating and Aggravating Factors of Migraine Versus Tension-type Headache', *Headache* 41, 6, 554–58, 2001.

10. Kelman, L., 'The Triggers or Precipitants of the Acute Migraine Attack', *Cephalalgia* 27, 5, 394–402, 2007.

11. Silva-Néto, R. P., 'Odorant Substances That Trigger Headaches in Migraine Patients', *Cephalalgia* 34, 1, 14–21, 2014.

12. Lima, A. M. *et al.*, 'Odors as Triggering and Worsening Factors for Migraine in Men', *Arquivos de Neuro-Psiquiatria* 69, 2B, 324–27, 2011.

13. 'The New South Wales Adult Health Survey 2002', *NSW Public Health Bulletin Supplement* 14, S-4, 81–82, 2003, health.nsw. gov.au/phb/Publications/NSW-adult-health-survey-2002.pdf quoted at 'Queensland Health Position Statement on Multiple Chemical Sensitivity', 2011, health.qld.gov.au/psu/docs/ pos-state-chemical.pdf

14. Caress, S. M. *et al.*, 'A National Population Study of the Prevalence of Multiple Chemical Sensitivity', *Archives of Environmental Health* 59, 6, 300–05, 2004 and Caress, S. M. *et al.*, 'Prevalence of Fragrance Sensitivity in the American Population', *Journal of Environmental Health* 71, 7, 46–50, 2009.

15. Steinemann, A., 'Fragranced Consumer Products: Exposures and Effects from Emissions', *Air Quality, Atmosphere & Health* 9, 8, 861–66, 2016.

16. Pain, 'When Others Abhor the Fragrance You Adore'.

17. Caress, S. M. *et al.*, 'National Prevalence of Asthma and Chemical Hypersensitivity: An Examination of Potential Overlap', *Journal of Occupational and Environmental Medicine* 47, 5, 518–22, 2005.

18. Caress, 'A National Population Study of the Prevalence of Multiple Chemical Sensitivity'.

19. Shim, C. *et al.*, 'Effects of Odors in Asthma', *American Journal of Medicine* 80, 1, 18–22, 1986.

20. Kumar, P. *et al.*, 'Inhalation Challenge Effects of Perfume Scent Strips in Patients with Asthma', *Annals of Allergy, Asthma & Immunology* 75, 5, 429–33, 1995.

21. Millqvist, E. *et al.*, 'Placebo-controlled Challenges with Perfume in Patients with Asthma-like Symptoms', *Allergy* 51, 6, 434–49, 1996.

22. Flegel, K. *et al.*, 'Artificial Scents Have No Place in Our Hospitals', *Canadian Medical Association Journal* 187, 16, 1187, 2015.

23. De Vader, C., 'Fragrance in the Workplace: What Managers Need to Know', *Journal of Management and Marketing Research* 3, 2010.

24. American Lung Association, 'Create a Lung Healthy Work Environment', lung.org/our-initiatives/healthy-air/indoor/at-work/guide-to-safe-and-healthy-workplaces/create-a-lung-healthy-work.html

25. Subbarao, P. *et al.*, 'Asthma: Epidemiology, Etiology and Risk Factors', *Canadian Medical Association Journal* 181, 9, 181–90, 2009.

26. Peiser, M. *et al.*, 'Allergic Contact Dermatitis: Epidemiology, Molecular Mechanisms, In Vitro Methods and Regulatory Aspects', *Cellular and Molecular Life Sciences* 69, 5, 763–81, 2012.

27. Jacob, S. *et al.*, 'Fragrances and Flavorants', *Dermatologist* 19, 7, 2011, the-dermatologist.com/content/fragrances-and-flavorants

28. Jacob, 'Fragrances and Flavorants'.

29. Bieber, T., 'Atopic Dermatitis', *Annals of Dermatology* 22, 2, 125–37, 2010.

30. Scheinman, P. L., 'Prevalence of Fragrance Allergy', *Dermatology* 205, 1, 98–102, 2002.

31. Zug, K. A. *et al.*, 'Patch-test Results of the North American Contact Dermatitis Group, 2005–2006', *Dermatitis* 20, 3, 149–60, 2009.

32. American Contact Dermatitis Society, 'ACDS Allergens of the Year', contactderm.org/14a/pages/index.cfm?pageid=3467

33. Uter, W. *et al.*, 'Categorization of Fragrance Contact Allergens for Prioritization of Preventive Measures: Clinical and Experimental Data and Consideration of Structure-activity Relationships', *Contact Dermatitis* 69, 4, 196–230, 2013.

CHAPTER 3: WHAT'S IN THE BOTTLE?

1. International Fragrance Association, 'Ingredients', ifraorg.org/en-us/ingredients#.V8asoijPbww
2. Süskind, P., *Perfume: The Story of a Murderer*, Alfred A. Knopf, 1986, p. 82.
3. Babu, G. *et al.*, 'Essential Oil Composition of Damask Rose (Rosa Damascena Mill.) Distilled Under Different Pressures and Temperatures', *Flavour and Fragrance Journal* 17, 2, 136–40, 2002.
4. Kuroda, K. *et al.*, 'Sedative Effects of the Jasmine Tea Odor and R-(-)linalool, One of its Major Odor Components, on Autonomic Nerve Activity and Mood States', *European Journal of Applied Physiology* 95, 2–3, 107–14, 2005.
5. Dietz, B. *et al.*, 'Botanical Dietary Supplements Gone Bad', *Chemical Research in Toxicology* 20, 4, 586–90, 2007.
6. Ngan, V., 'Balsam of Peru Allergy', DermNet New Zealand, 2002, dermnetnz.org/topics/balsam-of-peru-allergy
7. Henley, D. V. *et al.*, 'Prepubertal Gynecomastia Linked to Lavender and Tea Tree Oils', *New England Journal of Medicine* 356, 5, 479–85, 2007.
8. Memorial Sloan Kettering Cancer Center, 'Lavender', mskcc.org/cancer-care/integrative-medicine/herbs/lavender
9. International Fragrance Association, 'Standards Library', ifraorg.org/en-us/standards-library#.V8auZyjPbww
10. Quinessence Aromatherapy, 'Bulgarian Rose Otto', quinessence.com/bulgarian_rose_oil.htm
11. Davies, E., 'The Sweet Scent of Success', *Chemistry World*, 28 January 2009, chemistryworld.com/feature/the-sweet-scent-of-success/1012496.article

CHAPTER 4: WHAT NOSES KNOW

1. Havlíček, J. *et al.*, 'Non-advertized Does Not Mean Concealed: Body Odour Changes Across the Human Menstrual Cycle', *Ethology* 112, 1, 81–90, 2006.

2. Kuukasjärvi, S. *et al.*, 'Attractiveness of Women's Body Odors Over the Menstrual Cycle: The Role of Oral Contraceptives and Receiver Sex', *Behavioral Ecology* 15, 4, 579–84, 2004.

3. Weisfeld, G. E. *et al.*, 'Possible Olfaction-based Mechanisms in Human Kin Recognition and Inbreeding Avoidance', *Journal of Experimental Child Psychology* 85, 3, 279–95, 2003.

4. Kaitz, M. *et al.*, 'Mothers' Recognition of Their Newborns by Olfactory Cues', *Developmental Psychology* 20, 6, 587–91, 1987.

5. Vaglio, S., 'Chemical Communication and Mother–Infant Recognition', *Communicative & Integrative Biology* 2, 3, 279–81, 2009.

6. Nishitani, S. *et al.*, 'The Calming Effect of a Maternal Breast Milk Odor on the Human Newborn Infant', *Neuroscience Research* 63, 1, 66–71, 2009.

CHAPTER 5: BEHIND THE LABEL

1. Nazaroff, W. W. *et al.*, 'Indoor Air Chemistry: Cleaning Agents, Ozone and Toxic Air Contaminants', Final Report for the California Air Resources Board and California Environmental Protection Agency, 2006.

2. Steinemann, A. C. *et al.*, 'Fragranced Consumer Products: Chemicals Emitted, Ingredients Unlisted', *Environmental Impact Assessment Review* 31, 3, 328–33, 2011.

3. National Toxicology Program, 'Thirteenth Report on Carcinogens', 2014, ntp.niehs.nih.gov/pubhealth/roc/roc13/index.html

4. National Toxicology Program, 'Thirteenth Report on Carcinogens'.

5. Toxnet: Toxicology Data Network, 'Beta-pinene', toxnet.nlm.nih.gov/cgi-bin/sis/search/a?dbs+hsdb:@term+@DOCNO+5615 and World Health Organization, 'Concise International Chemical Assessment Document 5: Limonene', 1998, who.int/ipcs/publications/cicad/en/cicad05.pdf

6. Steinemann, 'Fragranced Consumer Products: Chemicals Emitted, Ingredients Unlisted'.

7. Steinemann, A. C., 'Volatile Emissions from Common Consumer Products', *Air Quality, Atmosphere & Health* 8, 3, 273–81, 2015.

8. Sutter County Superintendent of Schools, 'Safety Data Sheet: Denatured Alcohol', sutter.k12.ca.us/media/Facilities/MSDS/Denatured%20Alcohol%202.pdf

9. European Commission Scientific Committee on Consumer Safety, 'Opinion on Butylphenyl Methylpropional (BMHCA)', 16 March 2016, ec.europa.eu/health/scientific_committees/consumer_safety/docs/sccs_o_189.pdf

10. The Good Scents Company, 'Hydroxycitronellal', thegoodscents company.com/data/rw1000972.html

11. Toxnet: Toxicology Data Network, 'Geraniol', toxnet.nlm.nih.gov/cgi-bin/sis/search/a?dbs+hsdb:@term+@DOCNO+484

12. Toxnet: Toxicology Data Network, 'Benzyl Benzoate', toxnet.nlm.nih.gov/cgi-bin/sis/search/a?dbs+hsdb:@term+DOCNO+208

13. European Commission Scientific Committee on Consumer Safety, 'Perfume Allergies', 27 June 2012, ec.europa.eu/health/scientific_committees/opinions_layman/perfume-allergies/en/l-3/1-introduction.htm

14. International Fragrance Association, 'Standards Library'.

15. European Commission Scientific Committee on Consumer Safety, 'Perfume Allergies'.

16. Klammer, H. *et al.*, 'Multi-organic Risk Assessment of Estrogenic Properties of Octyl-methoxycinnamate in Vivo: A 5-day Sub-acute Pharmacodynamic Study with Ovariectomized Rats', *Toxicology* 215, 1–2, 90–96, 2005.

17. Charles, A. K. *et al.*, 'Oestrogenic Activity of Benzyl Salicylate, Benzyl Benzoate and Butylphenylmethylpropional (Lilial) in MCF7 Human Breast Cancer Cells In Vitro', *Journal of Applied Toxicology* 29, 5, 422–34, 2009.

18. Ministry of Environment and Food, Danish Environmental Protection Agency, 'Survey and Health Assessment of UV Filters', 2015, www2.mst.dk/Udgiv/publications/2015/10/978-87-93352-82-7.pdf

CHAPTER 6: WHO'S TESTING FRAGRANCE?

1. Australian Competition and Consumer Commission, 'Cosmetic Subscription Services Survey', 2015, productsafety.gov.au/system/files/Survey%20report%20-%20cosmetic%20subscription%20services%20compliance_0.pdf

2. Australian Government Department of Health, National Industrial Chemicals Notification and Assessment Scheme, nicnas.gov.au/regulation-and-compliance/nicnas-handbook/handbook-main-content/australian-inventory-of-chemical-substances/overview and nicnas.gov.au/chemical-information/imap-assessments/accelerated-assessment-of-industrial-chemicals-in-australia

3. Australian Government Department of Health, National Industrial Chemicals Notification and Assessment Scheme, nicnas.gov.au/chemical-information/imap-assessments/imap-assessments/tier-ii-environment-assessments/data-poor-fragrance-chemicals

4. Australian Competition and Consumer Commission, 'Research Survey of Formaldehyde in Cosmetics', 2010, productsafety.gov.au/publication/accc-research-survey-of-formaldehyde-in-cosmeticspdf

5. National Toxicology Program, 'Substances Listed in the Thirteenth Report on Carcinogens', ntp.niehs.nih.gov/ntp/roc/content/listed_substances_508.pdf

6. Australian Competition and Consumer Commission, 'Analytical Survey of Formaldehyde in False Eyelash Glues Supplied in Australia', 2015, productsafety.gov.au/publication/formaldehyde-in-false-eyelash-glues-supplied-in-australia

7. US Food and Drug Administration, 'FDA Authority Over Cosmetics: How Cosmetics Are Not FDA-approved, But Are FDA-regulated', fda.gov/Cosmetics/GuidanceRegulation/LawsRegulations/ucm074162.htm and 'Prohibited and Restricted Ingredients', fda.gov/Cosmetics/GuidanceRegulation/LawsRegulations/ucm127406.htm

8. European Commission Scientific Committee on Consumer Safety, 'Opinion on Fragrance Allergens in Cosmetic Products', 2011, ec.europa.eu/health/scientific_committees/consumer_safety/docs/sccs_o_073.pdf

9. European Commission, 'New Cosmetic Regulation to Strengthen Product Safety and to Cut Red Tape', press release, 5 February 2008, europa.eu/rapid/press-release_IP-08-184_en.htm

10. Patents, 'Substituted tetrahydronaphthalenes', US 2897237 A, google.com/patents/US2897237

11. Spencer, P. S. *et al.*, 'Neurotoxic Changes in Rats Exposed to the Fragrance Compound Acetyl Ethyl Tetramethyl Tetralin', *NeuroToxicology* 1, 1, 221–37, 1979 and Spencer, P. S. *et al.*, 'Neurotoxic fragrance produces ceroid and myelin disease', *Science* 204, 4393, 633–35, 1979.

12. Butterworth, K. R. *et al.*, 'Acute Toxicity of Thioguaiacol and of Versalide in Rodents', *Food and Cosmetics Toxicology* 19, 6, 753–55, 1982.

13. Akasaki, Y. *et al.*, 'Cerebellar Degeneration Induced by Acetyl-ethyl-tetramethyl-tetralin (AETT)', *Acta Neuropathologica* 80, 2, 129–37, 1990.

14. Koch-Henriksen, N. *et al.*, 'The Changing Demographic Pattern of Multiple Sclerosis Epidemiology', *Lancet Neurology* 9, 5, 520–32, 2010.

CHAPTER 7: IN DEFENCE OF FRAGRANCE

1. International Fragrance Association, ifraorg.org

2. International Fragrance Association, 'Standards', ifraorg.org/en-us/standards#.V9irr2WkwS4

3. International Fragrance Association, 'Standards Library'.

4. Tadeo, M., 'Iconic Chanel No 5 Perfume to Reformulate Under New EU Regulations', *Independent*, 29 May 2014, independent.co.uk/news/business/news/iconic-chanel-no-5-perfume-to-reformulate-under-new-eu-regulations-9451331.html

5. Comment, 18 November 2009, on Amelia, 'Is it Safe to Wear Old Perfume?', The Vintage Perfume Vault, 11 November 2009,

thevintageperfumevault.blogspot.com.au/2009/11/is-it-safe-to-wear-old-perfume.html

6. International Fragrance Association and Research Institute for Fragrance Materials, 'QRA Information Booklet Version 7.1', revised 9 July 2015, ifraorg.org/Upload/Download ButtonDocuments/c7b29dc8-19d2-4ffd-8aae-bb35ec2ae95b/ IFRA-RIFM%20QRA%20Information%20booklet%20V7.1%20 (July%209,%202015).pdf

7. International Fragrance Association, 'Standards Library'.

8. National Toxicology Program, 'Thirteenth Report on Carcinogens'.

9. International Fragrance Association, 'Standards Library'.

10. International Fragrance Association and Research Institute for Fragrance Materials, 'QRA Information Booklet Version 7.1'.

11. International Fragrance Association, 'Standards Library'.

12. International Fragrance Association and Research Institute for Fragrance Materials, 'QRA Information Booklet Version 7.1'.

13. International Fragrance Association and Research Institute for Fragrance Materials, 'QRA Information Booklet Version 7.1'.

CHAPTER 8: BEYOND THE LAB

1. University of Melbourne, *Up Close*, episode 341, upclose.unimelb. edu.au/episode/341-fume-view-consumer-products-and-your-indoor-air-quality

2. Saiyasombati, P. *et al.*, 'Two-stage Kinetic Analysis of Fragrance Evaporation and Absorption from Skin', *International Journal of Cosmetic Science* 25, 5, 235–43, 2003.

3. Shen, J. *et al.*, 'An *in silico* Skin Absorption Model for Fragrance Materials', *Food and Chemical Toxicology* 74, 164–76, 2014 at rifm.org/uploads/An%20in%20silico%20skin%20absorption %20model%20for%20fragrance%20materials%20FCT74%20 (10-2014)164-176.pdf

4. Wheeler, D. S. *et al.*, *Pediatric Critical Care Medicine: Basic Science and Clinical Evidence*, Springer, 2007, p. 1743.

5. Thyagarajan, A., 'New Guidelines Involving the Testing Done

on Animal Models', *Genetic Engineering & Biotechnology News*, 13 October 2015, genengnews.com/keywordsandtools/print/3/39557

6. International Fragrance Association, 'Compliance Programme', ifraorg.org/en-us/compliance-programme#.V8faSjPbww

7. Celeiro, M. *et al.*, 'Pressurized Liquid Extraction–Gas Chromatography–Mass Spectrometry Analysis of Fragrance Allergens, Musks, Phthalates and Preservatives in Baby Wipes', *Journal of Chromatography A* 1384, 9–21, 2015.

8. International Fragrance Association, 'Ingredients'.

CHAPTER 9: AFTERSHAVE WITH WHAT?

1. Turin, T., 'The Science of Scent', TED Talk, February 2005, ted.com/talks/luca_turin_on_the_science_of_scent?language=en

2. National Toxicology Program, Technical Report Series No. 422, 'Toxicology and Carcinogenesis Studies of Coumarin (CAS No. 91-64-5) in F344/N Rats and B6C3F1 Mice (Gavage Studies)', 1993, ntp.niehs.nih.gov/ntp/htdocs/lt_rpts/tr422.pdf

3. European Commission Scientific Committee on Food, 'Opinion on Coumarin', 22 September 1999, ec.europa.eu/food/fs/sc/scf/out40_en.pdf and Bundesinstitut für Risikobewertung, 'Consumers May Take in Larger Amounts of Coumarin from Cosmetics, Too', 20 December 2007, bfr.bund.de/cd/10569

4. Ford, R. A. *et al.*, 'The In Vivo Dermal Absorption and Metabolism of Coumarin by [4-14C] Rats and by Human Volunteers Under Simulated Conditions of Use in Fragrances', *Food and Chemical Toxicology* 39, 2, 153–62, 2001.

5. Abdo, K. M. *et al.*, 'Benzyl Acetate Carcinogenicity, Metabolism, and Disposition in Fischer 344 Rats and B6C3F1 Mice', *Toxicology* 37, 1–2, 159–70, 1985.

6. Cooper, S. D. *et al.*, 'The Identification of Polar Organic Compounds in Consumer Products and Their Toxicological Properties', *Journal of Exposure Analysis and Environmental Epidemiology* 5, 1, 57–75, 1995.

7. Cesta, M. F. *et al.*, 'Complex Histopathologic Response in

Rat Kidney to Oral Beta-myrcene: An Unusual Dose-related and Low-dose alpha2u-Globulin Nephropathy', *Toxicologic Pathology* 41, 8, 1068–77, 2013.

8. National Toxicology Program, Technical Report Series No. 557, 'Toxicology and Carcinogenesis Studies of Beta-myrcene in F344/N Rats and B6C3F1 Mice', 2010, ntp.niehs.nih.gov/ntp/htdocs/lt_rpts/tr557.pdf

9. Rastogi, S. C. *et al.*, 'Selected Important Fragrance Sensitizers in Perfumes—Current Exposures', *Contact Dermatitis* 56, 4, 201–04, 2007.

10. National Toxicology Program, Technical Report Series No. 551, 'Toxicology and Carcinogenesis Studies of Isoeugenol in F344/N Rats and B6C3F1 Mice', 2010, ntp.niehs.nih.gov/ntp/htdocs/lt_rpts/tr551.pdf

11. National Toxicology Program, Technical Report Series No. 253, 'Carcinogenesis Studies of Allyl Isovalerate (CAS No. 2835-39-4) in F344/N Rats and B6C3F1 Mice (Gavage Studies)', 1983, ntp.niehs.nih.gov/ntp/htdocs/lt_rpts/tr253.pdf

12. Bristol, D. W., 'NTP 3-month Toxicity Studies of Estragole (CAS No. 140-67-0) Administered by Gavage to F344/N Rats and B6C3F1 Mice', *National Toxicology Program Toxicology Report Series* 82, 1–111, 2011.

13. European Commission, 'Opinion of the Scientific Committee on Food on Estragole', 26 September 2001, ec.europa.eu/food/safety/docs/fs_food-improvement-agents_flavourings-out104.pdf

14. International Fragrance Association, 'Standards Library'.

15. International Agency for Research on Cancer, 'Monographs on the Evaluation of Carcinogenic Risks to Humans: Volume 71: Re-evaluation of Some Organic Chemicals, Hydrazine and Hydrogen Peroxide', 1999, monographs.iarc.fr/ENG/Monographs/vol71/mono71.pdf

16. National Toxicology Program, 'Thirteenth Report on Carcinogens'.

17. International Agency for Research on Cancer, 'Monographs on the Evaluation of Carcinogenic Risks to Humans: Volume 100F:

Formaldehyde', 2012, monographs.iarc.fr/ENG/Monographs/vol100F/mono100F-29.pdf

18. Nazaroff, 'Indoor Air Chemistry: Cleaning Agents, Ozone and Toxic Air Contaminants'.

19. de Groot, A. C. *et al.*, 'Formaldehyde-releasers: Relationship to Formaldehyde Contact Allergy: Contact Allergy to Formaldehyde and Inventory of Formaldehyde-releasers', *Contact Dermatitis* 61, 2, 63–85, 2009.

20. International Agency for Research on Cancer, 'Monographs on the Evaluation of Carcinogenic Risks to Humans: Volume 71: Re-evaluation of Some Organic Chemicals, Hydrazine and Hydrogen Peroxide'.

21. National Toxicology Program, 'Thirteenth Report on Carcinogens'.

22. European Commission Scientific Committee on Consumer Safety, 'Opinion on Acetaldehyde', 18 September 2012, ec.europa.eu/health/scientific_committees/consumer_safety/docs/sccs_o_104.pdf

23. Steinemann, 'Volatile Emissions from Common Consumer Products' and Black, R. E. *et al.*, 'Occurrence of 1,4-Dioxane in Cosmetic Raw Materials and Finished Cosmetic Products', *Journal of AOAC International* 84, 3, 666–70, 2001.

24. National Toxicology Program, 'Thirteenth Report on Carcinogens'.

25. National Industrial Chemicals Notification and Assessment Scheme, '1,4-Dioxane: Priority Existing Chemical No. 7: Full Public Report', 1998, available at clu-in.org/download/contaminantfocus/dioxane/PEC7_Full_Report_PDF.pdf

26. National Toxicology Program, 'Thirteenth Report on Carcinogens: Dichloromethane', ntp.niehs.nih.gov/ntp/roc/content/profiles/dichloromethane.pdf#search=Dichloromethane

27. Liu, T. *et al.*, 'Occupational Exposure to Methylene Chloride and Risk of Cancer: A Meta-analysis', *Cancer Causes & Control* 24, 12, 2037–49, 2013.

28. National Toxicology Program, 'Thirteenth Report on Carcinogens'.

29. International Fragrance Association, 'Standards Library'.

30. Johnson's, 'The Simple Truth: What's Changed', johnsonsbaby. com.au/difference/simple-truth
31. Thomas, K., 'The "No More Tears" Shampoo, Now with No Formaldehyde', *New York Times*, 17 January 2014, nytimes. com/2014/01/18/business/johnson-johnson-takes-first-step-in-removal-of-questionable-chemicals-from-products.html?_r=0
32. For example, *BBC News Magazine*, 'Is There a Danger from Scented Products?', 15 January 2016, bbc.com/news/magazine-35281338

CHAPTER 10: THE INDESTRUCTIBLES

1. Rimkus, G. G., 'Polycyclic Musk Fragrances in the Aquatic Environment', *Toxicology Letters* 111, 1–2, 37–56, 1999.
2. He, Y. J. *et al.*, 'Fate and Removal of Typical Pharmaceuticals and Personal Care Products by Three Different Treatment Processes', *Science of the Total Environment* 447, 248–54, 2013.
3. Klaschka, U. *et al.*, 'Occurrences and Potential Risks of 16 Fragrances in Five German Sewage Treatment Plants and Their Receiving Waters', *Environmental Science and Pollution Research International* 20, 4, 2456–71, 2013.
4. Waghulkar, V. M., 'Dark side of PPCP: An Unconscious Infiltration into Environment', *International Journal of ChemTech Research* 2, 2, 899–902, 2010.
5. Villa, S. *et al.*, 'Theoretical and Experimental Evidences of Medium Range Atmospheric Transport Processes of Polycyclic Musk Fragrances', *Science of the Total Environment* 481, 27–34, 2014.
6. Peters, R. J. *et al.*, 'Xeno-estrogenic Compounds in Precipitation', *Journal of Environmental Monitoring* 10, 760–69, 2008.
7. Macherius, A. *et al.*, 'Uptake of Galaxolide, Tonalide, and Triclosan by Carrot, Barley and Meadow Fescue Plants', *Journal of Agricultural and Food Chemistry* 60, 32, 7785–91, 2012.
8. Reiner, J. L. *et al.*, 'Polycyclic Musks in Water, Sediment, and Fishes from the Upper Hudson River, New York, USA', *Water, Air, & Soil Pollution* 214, 1, 335–42, 2011.

9. Schiavone, A. *et al.*, 'Polybrominated Diphenyl Ethers, Polychlorinated Naphthalenes and Polycyclic Musks in Human Fat from Italy: Comparison to Polychlorinated Biphenyls and Organochlorine Pesticides', *Environmental Pollution* 158, 2, 599–606, 2010.

10. Hutter, H. P. *et al.*, 'Higher Blood Concentrations of Synthetic Musks in Women Above Fifty Years Than in Younger Women', *International Journal of Hygiene and Environmental Health* 213, 2, 124–30, 2010.

11. Lu, Y. *et al.*, 'Occurrence of Synthetic Musks in Indoor Dust from China and Implications for Human Exposure', *Archives of Environmental Contamination and Toxicology*, 60, 1, 182–89, 2011.

12. Kubwabo, C. *et al.*, 'Determination of Synthetic Musk Compounds in Indoor House Dust by Gas Chromatography–Ion Trap Mass Spectrometry', *Analytical and Bioanalytical Chemistry* 404, 2, 467–77, 2012.

13. Sathyanarayana, S. *et al.*, 'Baby Care Products: Possible Sources of Infant Phthalate Exposure', *Pediatrics* 121, 2, 260–68, 2008.

14. Taylor, K. M. *et al.*, 'Human Exposure to Nitro Musks and the Evaluation of Their Potential Toxicity: An Overview', *Environmental Health* 13, 1, 14, 2014.

15. Lignell, S. *et al.*, 'Temporal Trends of Synthetic Musk Compounds in Mother's Milk and Associations with Personal Use of Perfumed Products', *Environmental Science & Technology* 42, 17, 6743–48, 2008.

16. Mori, T. *et al.*, 'Hormonal Activity of Polycyclic Musks Evaluated by Reporter Gene Assay', *Environmental Science* 14, 4, 195–202, 2007.

17. Bitsch, N. *et al.*, 'Estrogenic Activity of Musk Fragrances Detected by the E-screen Assay Using Human mcf-7 Cells', *Archives of Environmental Contamination and Toxicology* 43, 3, 257–64, 2002.

18. Lange, C. *et al.*, 'Estrogenic Activity of Constituents of Underarm

Deodorants Determined by E-Screen Assay', *Chemosphere* 108, 101–06, 2014.

19. Schreurs, R. H. *et al.*, 'In Vitro and In Vivo Antiestrogenic Effects of Polycyclic Musks in Zebrafish', *Environmental Science & Technology* 38, 4, 997–1002, 2004.

20. Schreurs, R. H. *et al.*, 'Interaction of Polycyclic Musks and UV Filters with the Estrogen Receptor (ER), Androgen Receptor (AR), and Progesterone Receptor (PR) in Reporter Gene Bioassays', *Toxicological Sciences* 83, 2, 264–72, 2005.

21. Li, Z. *et al.*, 'Effects of Polycyclic Musks HHCB and AHTN on Steroidogenesis in H295R Cells', *Chemosphere* 90, 3, 1227–35, 2013.

22. Schreurs, R. H. *et al.*, 'Transcriptional Activation of Estrogen Receptor ERalpha and ERbeta by Polycyclic Musks is Cell Type Dependent', *Toxicology and Applied Pharmacology* 183, 1, 1–9, 2002.

23. Schlumpf, M. *et al.*, 'Endocrine Activity and Developmental Toxicity of Cosmetic UV Filters—An Update', *Toxicology* 205, 1–2, 113–22, 2004.

24. Kumar, N. *et al.*, 'Assessment of Estrogenic Potential of Diethyl Phthalate in Female Reproductive System Involving Both Genomic and Non-genomic Actions', *Reproductive Toxicology* 49, 12–26, 2014.

25. Schreurs, 'Interaction of Polycyclic Musks and UV Filters with the Estrogen Receptor (ER), Androgen Receptor (AR), and Progesterone Receptor (PR) in Reporter Gene Bioassays'.

26. Simmons, D. B. *et al.*, 'Interaction of Galaxolide® with the Human and Trout Estrogen Receptor-alpha', *Science of the Total Environment* 408, 24, 6158–64, 2010.

27. van der Burg, B. *et al.*, 'Endocrine Effects of Polycyclic Musks: Do We Smell a Rat?', *International Journal of Andrology* 31, 2, 188–93, 2008.

28. van Meeuwen, J. A., 'Aromatase Inhibiting and Combined Estrogenic Effects of Parabens and Estrogenic Effects of Other Additives in Cosmetics', *Toxicology and Applied Pharmacology* 230, 3, 372–82, 2008.

29. Quoted in Freinkel, S., *Plastic: A Toxic Love Story*, Text Publishing, 2011, p. 94.

CHAPTER 11: UNINTENDED CONSEQUENCES

1. '7.1 Invertebrates' in Kortenkamp, A. *et al.*, *State of the Art Assessment of Endocrine Disrupters: Final Report: Annex 1*, 29 January 2012, ec.europa.eu/environment/chemicals/endocrine/pdf/annex1_summary_state_of_science.pdf

2. Foster, P. M., 'Disruption of Reproductive Development in Male Rat Offspring Following In Utero Exposure to Phthalate Esters', *International Journal of Andrology* 29, 1, 140–47, 2006.

3. '4.1 Male Reproductive Health' in Kortenkamp, *State of the Art Assessment of Endocrine Disrupters: Final Report: Annex 1*.

4. Virtanen, H. E. *et al.*, 'Cryptorchidism and Endocrine Disrupting Chemicals', *Molecular and Cellular Endocrinology* 355, 2, 208–20, 2012.

5. '4.1 Male Reproductive Health' in Kortenkamp, *State of the Art Assessment of Endocrine Disrupters: Final Report: Annex 1*.

6. '4.1 Male Reproductive Health' in Kortenkamp, *State of the Art Assessment of Endocrine Disrupters: Final Report: Annex 1*.

7. Eisenhardt, S. *et al.*, 'Nitromusk Compounds in Women with Gynecological and Endocrine Dysfunction', *Environmental Research* 87, 3, 123–130, 2001.

8. Virtanen, 'Cryptorchidism and Endocrine Disrupting Chemicals'.

9. Klip, H. *et al.*, 'Hypospadias in Sons of Women Exposed to Diethylstilbestrol In Utero: A Cohort Study', *Lancet* 359, 9312, 1102–07, 2002.

10. Kalfa, N. *et al.*, 'Prevalence of Hypospadias in Grandsons of Women Exposed to Diethylstilbestrol During Pregnancy: A Multigenerational National Cohort Study', *Fertility and Sterility* 95, 8, 2574–77, 2011.

11. Swan, S. H. *et al.*, 'Decrease in Anogenital Distance Among Male Infants with Prenatal Phthalate Exposure', *Environmental Health Perspectives* 113, 8, 1056–61, 2005.

12. Main, K. M. *et al.*, 'Human Breast Milk Contamination

with Phthalates and Alterations of Endogenous Reproductive Hormones in Infants Three Months of Age', *Environmental Health Perspectives* 114, 2, 270–76, 2006.

13. Sharpe, R. M. *et al.*, 'Are Oestrogens Involved in Falling Sperm Counts and Disorders of the Male Reproductive Tract?', *Lancet*, 341, 8857, 1392–95, 1993 and Skakkebaek, N. E. *et al.*, 'Testicular Dysgenesis Syndrome: An Increasingly Common Developmental Disorder with Environmental Aspects', *Human Reproduction* 16, 5, 972–78, 2001.

14. Huang, L. *et al.*, 'Estrogenic Regulation of Signaling Pathways and Homeobox Genes During Rat Prostate Development', *Journal of Andrology* 25, 3, 330–37, 2004.

15. vom Saal, F. S. *et al.*, 'Prostate Enlargement in Mice Due to Fetal Exposure to Low Doses of Estradiol or Diethylstilbestrol and Opposite Effects at High Doses', *Proceedings of the National Academy of Sciences of the United States of America* 94, 5, 2056–61, 1997.

16. Cancer Australia, 'Prostate Cancer Statistics', prostate-cancer. canceraustralia.gov.au/statistics

17. Cancer Australia, 'Breast Cancer in Australia', canceraustralia. gov.au/affected-cancer/cancer-types/breast-cancer/breast-cancer-statistics

18. Breastcancer.org, 'Hormone Receptor Status', breastcancer.org/ symptoms/diagnosis/hormone_status

19. '5.1 Breast Cancer' in Kortenkamp, *State of the Art Assessment of Endocrine Disrupters: Final Report: Annex 1.*

20. Palmer, J. R. *et al.*, 'Prenatal Diethylstilbestrol Exposure and Risk of Breast Cancer', *Cancer Epidemiology Biomarkers and Prevention* 15, 8, 1509–14, 2006.

21. Centers for Disease Control and Prevention, 'Information to Identify and Manage DES Patients', cdc.gov/des/hcp/information/ daughters/risks_daughters.html

22. Rossouw, J. E. *et al.*, 'Risks and Benefits of Estrogen Plus Progestin in Healthy Postmenopausal Women: Principal Results from the Women's Health Initiative Randomized Controlled

Trial', *Journal of the American Medical Association* 288, 3, 321–33, 2002.

23. Beral, V. *et al.*, 'Breast Cancer and Hormone-Replacement Therapy in the Million Women Study', *Lancet* 362, 9382, 419–427, 2003.

24. Verkooijen, H. M. *et al.*, 'The Incidence of Breast Cancer and Changes in the Use of Hormone Replacement Therapy: A Review of the Evidence', *Maturitas* 64, 2, 80–85, 2009.

25. '5.1 Breast Cancer' in Kortenkamp, *State of the Art Assessment of Endocrine Disrupters: Final Report: Annex 1*.

26. '5.1 Breast Cancer' in Kortenkamp, *State of the Art Assessment of Endocrine Disrupters: Final Report: Annex 1*.

27. '5.1 Breast Cancer' in Kortenkamp, *State of the Art Assessment of Endocrine Disrupters: Final Report: Annex 1*.

28. Depue, R. H. *et al.*, 'Estrogen Exposure During Gestation and Risk of Testicular Cancer', *Journal of the National Cancer Institute* 71, 6, 1151–55, 1983.

29. Bouskine, A. *et al.*, 'Estrogens Promote Human Testicular Germ Cell Cancer Through a Membrane-mediated Activation of Extracellular Regulated Kinase and Protein Kinase A', *Endocrinology* 149, 2, 565–73, 2007.

30. '4.1 Male Reproductive Health' in Kortenkamp, *State of the Art Assessment of Endocrine Disrupters: Final Report: Annex 1*.

31. Kortenkamp, *State of the Art Assessment of Endocrine Disrupters: Final Report: Annex 1*, pp. 135, 169, 181, 198, 218, 292–93.

32. '5.2 Prostate Cancer' in Kortenkamp, *State of the Art Assessment of Endocrine Disrupters: Final Report: Annex 1*.

33. '5.3 Testis Cancer' in Kortenkamp, *State of the Art Assessment of Endocrine Disrupters: Final Report: Annex 1* and Huyghe, E. *et al.*, 'Increasing Incidence of Testicular Cancer Worldwide: A Review', *Journal of Urology* 170, 1, 5–11, 2003.

34. Sharpe, R., 'Male Reproductive Health Disorders and the Potential Role of Exposure to Environmental Chemicals', CHEM Trust, chemtrust.org.uk/wp-content/uploads/ProfRSHARPE-MaleReproductiveHealth-CHEMTrust09-1.pdf

35. Main, K. M. *et al.*, 'Genital Anomalies in Boys and the Environment', *Best Practice & Research Clinical Endocrinology & Metabolism* 24, 2, 279–89, 2010.

36. Nassar, N. *et al.*, 'Increasing Prevalence of Hypospadias in Western Australia, 1980–2000', *Archives of Disease in Childhood* 92, 7, 580–84, 2007.

37. Cancer Australia, 'Breast Cancer in Australia', canceraustralia. gov.au/affected-cancer/cancer-types/breast-cancer/breast-cancer-statistics

38. Li, C. I. *et al.*, 'Incidence of Invasive Breast Cancer by Hormone Receptor Status from 1992 to 1998', *Journal of Clinical Oncology* 21, 1, 28–34, 2003.

39. '4.2 Female Precocious Puberty' in Kortenkamp, *State of the Art Assessment of Endocrine Disrupters: Final Report: Annex 1*.

40. '5.5 Other Hormonal Cancers: Ovarian and Endometrial Cancers' in Kortenkamp, *State of the Art Assessment of Endocrine Disrupters: Final Report: Annex 1*.

41. Kortenkamp, *State of the Art Assessment of Endocrine Disrupters: Final Report: Annex 1*.

42. 'Report to the U.S. Consumer Product Safety Commission by the Chronic Hazard Advisory Panel on Phthalates and Phthalate Alternatives', July 2014, cpsc.gov/PageFiles/169902/CHAP-REPORT-With-Appendices.pdf

43. Body Shop, 'Chemicals Strategy, July 2008', thebodyshop.ca/en/pdfs/values-campaigns/BSI_Chemicals_Strategy.pdf

44. Johnson's, 'The Simple Truth: Answers to Popular Questions', johnsonsbaby.com.au/difference/simple-truth

CHAPTER 12: SHARING THE AIR

1. See Steinemann, A., 'Safety Management: Indoor Environmental Quality Policy', drsteinemann.com/Resources/CDC%20Indoor%20Environmental%20Quality%20Policy.pdf

2. US Equal Employment Opportunity Commission, transcript of Town Hall Listening Session, Chicago, 17 November 2009, eeoc.gov/eeoc/events/transcript-chic.cfm and Peeples, L., 'Chemically

Sensitive Find Sanctuary in Fragrance-free Churches', *Huffington Post*, 27 October 2013, huffingtonpost.com.au/entry/chemical-sensitivity-fragrance-church_n_4163785

3. Steinemann, 'Fragranced Consumer Products: Exposures and Effects from Emissions'.

4. Capital Health Cancer Care Program, 'Cancer Care: A Guide for Patients, Families and Caregivers', cancercare.ns.ca/site-cc/media/cancercare/CCNS_CH_Guide_05_11(1).pdf

5. Garcia, M., 'Scents and Sensitivity: Considering Passenger Allergies to Fragrance', APEX, 25 January 2016, apex.aero/2016/01/22/passenger-allergies-to-fragrance

6. Canadian Human Rights Commission, 'Policy on Environmental Sensitivities', January 2014, chrc-ccdp.gc.ca/sites/default/files/policy_sensitivity_0.pdf

7. Steinemann, 'Safety Management: Indoor Environmental Quality Policy'.

8. Massachusetts Nurses Association, 'Model for a Fragrance-free Policy', 15 April 2016, massnurses.org/health-and-safety/articles/chemical-exposures/p/openItem/1346#model

9. Safe Work Australia, 'Managing Risks of Airborne Contaminants', safeworkaustralia.gov.au/sites/swa/whs-information/hazardous-chemicals/airborne-contaminants/pages/managing-risks-airborne-contaminants

10. Safe Work Australia, 'Managing Risks of Hazardous Chemicals in the Workplace: Code of Practice', July 2012, safeworkaustralia.gov.au/sites/SWA/about/Publications/Documents/697/Managing%20Risks%20of%20Hazardous%20Chemicals2.pdf

11. Australian Human Rights Commission, 'Use of Chemicals and Materials' and 'Action Plans' in 'Access: Guidelines and Information', humanrights.gov.au/publications/access-guidelines-and-information

12. WorkplaceOHS, 'Can We Enforce a "No Personal Spray" Policy?', 25 February 2008, workplaceohs.com.au/risk-management/roles-responsibilities/q-a/can-we-enforce-a-no-personal-spray-policy

CHAPTER 13: THE OPPOSITE OF FRAGRANCE

1. Symphony Nova Scotia, 'Concerts FAQs', symphonynovascotia. ca/faqs/concert
2. Turin, L., *et al*, *Perfumes: The A–Z Guide*, Penguin USA, 2008, p. 17.
3. Turin, *Perfumes*, p. 15.
4. Turin, *Perfumes*, p. 290.
5. 'Scent Branding Sweeps the Fragrance Industry', *Bloomberg Businessweek*, 17 June 2010, bloomberg.com/news/articles/ 2010-06-16/scent-branding-sweeps-the-fragrance-industry
6. Schifferstein, H. *et al.*, 'The Signal Function of Thematically (In)congruent Ambient Scents in a Retail Environment', *Chemical Senses* 27, 6, 539–49, 2002.

ACKNOWLEDGMENTS

This was an odd and difficult book to write, and I'm grateful to those early readers who offered support. First among them are Tom and Alice Petty—your thoughtful reading and imaginative suggestions made it all seem not just possible but worth doing—thank you.

Several readers of early drafts gave the book the benefit of their expertise and insight. Engaging with a draft of someone else's book is an act of great generosity and goodwill, which I deeply appreciate. Others offered encouragement, information and ideas without which the book would have foundered and sunk at an early stage. Warmest thanks to you all: Margaret Armstrong, Gail Bell, Helena Berenson, Henry Berenson, Susan Hampton, Ashley Hay, Ralph Higgins, Geoff Holden, Carole Hungerford,

Deborah Kingsland, Catherine McIver, Col Madden, Helen Maxwell, Ellen Moore, Clare Pain, Bruce Petty, Carolyn Re, Loretta Re, Anna Rigg, Julie Rigg, Anne Steinemann, Chris Wallace, and Sarah Wilson.

My agent Barbara Mobbs was, as always, a source of good sense and warm support, and the team at Text Publishing gave me the best professional backing any author could wish for. Special thanks to Jane Novak, who (I learned after our book tour together) even thought about buying me a roll of packaging tape, but was afraid I'd think she was crazy.

All around the world, researchers labour to deepen our understanding of fragrance chemicals and their effects on our health. Their work is a gift for all of us who want to know more. I offer them my respectful and admiring gratitude.